Leading Effective Engineering Teams

Lessons for Individual Contributors and Managers from 10 Years at Google

Addy Osmani

Beijing • Boston • Farnham • Sebastopol • Tokyo

Leading Effective Engineering Teams

by Addy Osmani

Published by O'Reilly Media, Inc., 1005 Gravenstein Highway North, Sebastopol, CA 95472.

O'Reilly books may be purchased for educational, business, or sales promotional use. Online editions are also available for most titles (*https://oreilly.com*). For more information, contact our corporate/institutional sales department: 800-998-9938 or *corporate@oreilly.com*.

Acquisitions Editor: David Michelson
Development Editor: Rita Fernando
Production Editor: Kristen Brown
Copyeditor: nSight, Inc.
Proofreader: Doug McNair
Indexer: WordCo Indexing Services, Inc.
Interior Designer: Monica Kamsvaag
Cover Designer: Susan Thompson
Illustrator: Kate Dullea

June 2024: First Edition

Revision History for the First Edition

2024-06-11: First Release

See *http://oreilly.com/catalog/errata.csp?isbn=9781098148249* for release details.

978-1-098-14824-9

[LSI]

Contents

Foreword

When I transitioned from a software engineer to an engineering manager some time ago, I was surprisingly unprepared on what to do. I looked to my manager and peer engineering managers for advice and guidance, and I followed a trial-and-error approach to figure out what works. I had hoped that I'd do more "good" than "bad," correct the "bad" quickly enough, and not make too many hard-to-fix mistakes.

I wish I could say that times have changed and engineering managers these days can rely on more training or have access to better resources as they seek to become great leaders. However, most managers still do what I did: try various approaches and stick with ones that seem to be working. It's little wonder that the manager job feels as difficult as it does—or that burnout is still high among managers.

One company that decided to try to change how it supported engineering managers in their growth path was Google. Instead of relying on anecdotes and gut feelings, Google decided to quantify and verbalize the qualities that made great engineering managers. To this end, it ran two large-scale research projects, both of them the first of such scale. In 2008, Google ran Project Oxygen and identified eight attributes of good managers. In 2012, the company ran Project Aristotle and identified five factors that contribute to the success of software engineering teams.

Addy Osmani joined Google in the same year that Project Aristotle was kicked off. Working inside a company that is obsessed with understanding what makes great engineering leaders—and then applying these lessons to make them better managers—is a helpful place to transition into management. Addy made this switch more than a decade ago. Since then, Addy has been taking notes on what kept him effective as a leader, and he is finally sharing these in this book.

The challenge of being a great engineering leader is that there's so much advice out there, not to mention the dozens of frameworks and mental models. How do you reconcile all these? In the first part of this book, Addy walks through the "modern theory" of engineering management: research and mental models on this space that have stood the test of time for useful management practices—ones that work well for *tech* teams. Many of these models are complementary, and some slightly contradict one another. In your own journey, you'll need to pick and choose the ones that are most helpful in your specific context—and you will likely find yourself changing philosophies as you change teams or companies.

The craft of leadership is far more practical than theoretical, though. The second part of this book is where Addy shares his practical approach to leading engineering teams, and this is the part where Addy shares a condensed history of a decade of leading tech teams. How should we deal with a team that addresses most issues heroically—and hastily—*just* before the release? What about a team where most members "rubber-stamp" PRs (pull requests) or have unusually long-running ones? And should you take action if your team neglects to do any form of retrospective—or is this something that can safely be ignored?

In addition to sharing his own, distilled experience, Addy allows us to peek at how effective managers at Google operate, as well as what effective leadership looks like in other environments like startups or large enterprises.

In the end, don't lose sight of what makes a good leader: they achieve organizational goals. But what makes a great leader? They do it in a way that inspires and motivates the team. And how do they do all of this? Before I read this book, my answer would have been that they learn how to do it by shadowing a standout engineering leader. Now, you can get distilled insights from such a leader by carrying on reading this book—and I hope you will enjoy it as much as I did.

—Gergely Orosz
Former engineering manager at Uber,
author of The Pragmatic Engineer
Amsterdam, March 2024

Preface

Today, the demands placed on engineering teams are more complex and multi-faceted than ever before, and the ability to scale your effectiveness as a leader has emerged as a critical skill. Regardless of whether you are at the helm of a small, nimble startup team or spearheading large-scale initiatives within a global corporation, the core principles of fostering trust, cultivating a growth mindset, and driving accountability remain universal and essential to achieving success.

Throughout my long tenure at Google, I have been humbled to have the extraordinary privilege of leading engineering teams in a wide range of capacities. From my early days as an individual contributor, when I focused on mentoring and nurturing the growth of my fellow engineers, to my roles as a tech lead manager and org leader at various levels, I have accumulated a wealth of knowledge and insights that have fundamentally shaped my understanding of what it takes to build and lead truly exceptional engineering teams. This book represents a distillation of those invaluable lessons.

This book is for engineers wanting to move into leadership roles or engineering leaders who want evidence-based guidance to improve their effectiveness and that of their teams. It is a comprehensive guide to the strategies, frameworks, and best practices that I have found to be most effective in unlocking the full potential of engineering teams and driving transformative results. By sharing real-world examples, practical insights, and actionable advice, I aim to empower you with the tools and knowledge you need to become an exceptional engineering leader in your own right.

At the heart of this book lies a deep exploration of the key traits and behaviors that distinguish highly effective engineers and engineering leaders from their peers. These are the individuals who consistently deliver outstanding results, inspire their teams to reach new heights, and make a lasting impact on the projects and initiatives they lead. By understanding and embodying these characteristics, you, too, can set yourself apart and make a meaningful difference in your role.

One of the central themes that permeates every chapter of this book is the paramount importance of creating a culture of psychological safety within your engineering team. In a psychologically safe environment, team members feel empowered to take risks, voice their ideas and opinions openly, and view failures as valuable opportunities for growth and learning. As a leader, it is your responsibility to actively foster and maintain this type of culture, where experimentation, innovation, and continuous improvement are not only encouraged but celebrated as essential components of success.

Throughout the book, you will dive deep into the dynamics of high-performing engineering teams, examining the strategies and practices that enable them to consistently deliver exceptional results. From the learning about the art of effective communication and collaboration to exploring the implementation of robust processes for decision making and problem-solving, you will gain a comprehensive understanding of the key ingredients that go into building and sustaining a team that operates at the highest level.

But the path to becoming an outstanding engineering leader extends far beyond gaining technical expertise and project management skills. It requires a multifaceted approach that encompasses gaining self-awareness, cultivating emotional intelligence, and developing the ability to deftly navigate complex interpersonal dynamics. In the chapters that follow, you will explore powerful techniques for self-advocacy and personal growth, helping you to amplify your impact and that of your team within the broader organizational landscape.

Drawing upon the wealth of knowledge shared within these pages, you will learn how to cultivate a deep sense of trust, commitment, and accountability among your team members. You will delve into the intricacies of providing constructive feedback, setting clear goals and expectations, and empowering your team to take ownership of their work and their professional development. By mastering these skills, you will be well equipped to create a team culture that fosters high levels of productivity and is deeply fulfilling for all involved.

Whether you are an aspiring engineering leader just embarking on your journey or a seasoned veteran looking to take your team's performance to new heights, this book will serve as an indispensable resource, providing you with the insights, tools, and frameworks needed to achieve your goals. By leveraging the collective wisdom and experiences shared within these pages, you will be able to drive meaningful change, overcome obstacles, and attain extraordinary results.

In the following chapters, you will find practical exercises, thought-provoking questions, and actionable strategies that you can begin implementing immediately within your own team and organization. I invite you to fully engage with these materials, reflect deeply on your own experiences and challenges, and use the insights gained to chart a course toward greater effectiveness, efficiency, and impact.

As we delve into the art and science of leading extraordinary engineering teams, it is important to remember that this is not a journey that you have to undertake alone. I will be sharing my personal experiences and insights as well as the wisdom and advice of other accomplished engineering leaders who have graciously contributed their knowledge to this project.

In addition to the wealth of practical advice and real-world examples contained within these pages, I have included a series of in-depth case studies that showcase the transformative power of effective engineering leadership in action. These case studies feature some of the most innovative and successful engineering teams from across the industry, offering valuable insights into how they have overcome challenges, driven innovation, and achieved remarkable results.

By studying these real-world examples and learning from the experiences of other exceptional engineering leaders, you will gain a deeper understanding of what it takes to build and lead high-performing teams in today's complex technological landscape. Whether you are looking to improve your team's productivity, enhance collaboration and communication, or drive innovation and growth, these case studies will provide you with the inspiration and practical guidance you need to achieve your goals.

By tapping into this collective well of expertise and experience, you will be able to learn from the successes and challenges of others and apply those lessons to your own unique context and circumstances. Whether you are seeking to build stronger relationships with your team members, improve your communication and collaboration skills, or develop a more strategic and visionary approach to leadership, this book will serve as a valuable companion and guide on your journey.

As we set out on this journey together, I encourage you to approach each chapter with an open mind and a willingness to challenge your preconceptions. The path to becoming an exceptional engineering leader is one of continuous learning, adaptation, and growth. It requires a steadfast commitment to your own development as well as to the growth and success of those around you. Let's begin.

O'Reilly Online Learning

O'REILLY®

For more than 40 years, *O'Reilly Media* has provided technology and business training, knowledge, and insight to help companies succeed.

Our unique network of experts and innovators share their knowledge and expertise through books, articles, and our online learning platform. O'Reilly's online learning platform gives you on-demand access to live training courses, in-depth learning paths, interactive coding environments, and a vast collection of text and video from O'Reilly and 200+ other publishers. For more information, visit *https://oreilly.com.*

How to Contact Us

Please address comments and questions concerning this book to the publisher:

O'Reilly Media, Inc.

1005 Gravenstein Highway North

Sebastopol, CA 95472

800-889-8969 (in the United States or Canada)

707-827-7019 (international or local)

707-829-0104 (fax)

support@oreilly.com

https://www.oreilly.com/about/contact.html

We have a web page for this book, where we list errata, examples, and any additional information. You can access this page at *https://oreil.ly/leadEffEng-Teams.*

For news and information about our books and courses, visit *https://oreilly.com.*

Find us on LinkedIn: *https://linkedin.com/company/oreilly-media.*

Watch us on YouTube: *https://youtube.com/oreillymedia.*

Acknowledgments

Writing this book has been an incredible journey, one that would not have been possible without the unwavering support, guidance, and contributions of numerous folks who have generously shared their time, expertise, and insights throughout the process.

First and foremost, I would like to extend my deepest gratitude to the exceptional tech reviewers who have played an instrumental role in shaping the content and ensuring the accuracy and relevance of the material presented within these pages. Kate Wardin, Francisco Trindade, Maxi Ferreira, and Sriram Krishnan, your invaluable feedback, keen observations, and wealth of experience have been essential in helping me refine the ideas and strategies outlined in this book. Your dedication to providing thoughtful and constructive critiques has elevated the quality of this work immeasurably, and for that, I am truly grateful.

I would also like to express my heartfelt appreciation to the outstanding editors who have been by my side every step of the way. Rita Fernando and Leena Sohoni-Kasture, your tireless efforts, meticulous attention to detail, and unwavering commitment to excellence have been nothing short of remarkable. Your guidance, patience, and editorial expertise have been instrumental in bringing this book to life, and I am deeply indebted to you for your unwavering support and dedication.

Finally, I would like to express my profound gratitude to my family, friends, and fellow Googlers, whose support and understanding have been the bedrock upon which this entire endeavor has been built. Your unwavering belief in the project and your constant encouragement have been the driving force behind this book's completion.

To all those who have contributed to the creation of this book, whether directly or indirectly, I extend my deepest appreciation and gratitude. Your collective efforts and support have made this project a reality, and I am honored to have had the opportunity to collaborate with such an exceptional group of individuals.

What Makes a Software Engineering Team Effective?

Some teams seem to operate like well-oiled machines, churning out successes. Communication flows seamlessly, they meet deadlines with a smile, and they tackle challenges head-on. Conversely, other teams struggle to reach every milestone. Communication is chaotic, and meeting deadlines is a challenge. What makes the successful teams effective? It's usually a mix of things: clear plans, honest talk, a healthy dose of trust, and a shared belief in what they're doing. Some teams already have the rhythm and the steps down pat, while others are still figuring things out. But the good news is that everyone can learn the steps. Even the most stumbling crew can find its rhythm with a little practice.

This rhythm manifests itself in software engineering teams as their ability to produce useful products or product features by writing code, testing it, and releasing it to the world. Teams that do this regularly are said to be *effective*. So, to build great software, we must first build effective engineering teams.

Throughout my 25+ years of experience leading engineering teams at Google and other tech companies, I've seen firsthand how team dynamics can make or break a project. Building effective teams is not just about assembling the right technical skills; it's about fostering a culture of collaboration, trust, and shared purpose. In this chapter, I'll share some of the key lessons I've learned about what makes engineering teams successful, drawing on both research and my own experience in the trenches.

What makes an engineering team effective hinges on the key thing that distinguishes teams from groups. On the one hand, a *group* is a collection of individuals who coordinate their efforts. On the other hand, a *team* is a group that is bound by shared responsibilities and goals. Their members work together and share mutual accountability to solve problems and achieve common goals. When teams plan their work, review progress, or make decisions, they consider the skills and availability of all the members and not just those of one individual. This shared goal is what drives an effective team.

I have had the opportunity to observe or be a part of such teams at Google. These teams are passionate about achieving their goals. They find brainstorming sessions fun rather than stressful. Team members may write and test code on their respective machines, but they are collectively tuned in to a unified vision of what the code should achieve. There have been times when they had to resolve some difficult issues, but a culture of collaboration, innovation, and mutual respect helped to see them through such times.

Leaders are an important part of this picture. As a software engineering leader who wishes to make your team effective, you serve as an anchor that connects individual team members to the shared responsibilities and goals of the team. You provide the vision, direction, guidance, and environmental framework necessary to form this connection.

Although it's possible to have a team without a leader, the team will go much further with the support of a good leader—and that's where you come in!

Building an effective software engineering team takes work. Many factors can influence the success of a software engineering team, such as team composition, communication, leadership, and work processes. This chapter will explore what traits make teams effective and how to build them into your team. These traits will be things you can look for when hiring, but they're also traits you can nurture in your existing team.

Research on What Makes Teams Effective

First, let's examine what makes teams effective. To do so, let us look at some of the extensive research that has already been done on this topic.

PROJECT ARISTOTLE

Google conducted one of the best-known studies on effective software engineering teams, known as Project Aristotle (*https://oreil.ly/N3iky*).[1] The project aimed to identify the factors that make some teams more successful than others. The study was based on the premise that the composition of a team was not the most critical factor in determining success but rather how team members interacted with each other.

> **Note**
>
> Before Project Aristotle, there was Project Oxygen, which looked into what traits make for a good manager. Some of the insights in this chapter were informed by the results of Project Oxygen, which I'll talk about in detail in Chapter 4.

To determine what makes teams effective, the researchers first had to define what *effectiveness* means and how to measure it. They noticed that different roles had different perspectives on effectiveness. In general, whereas executives were interested in results such as sales numbers or product launches, team members thought that team culture was the key to team effectiveness. The team leaders indicated that ownership, vision, and goals were the most important measures.

Eventually, the researchers decided to study certain qualitative and quantitative factors that might impact team effectiveness, such as the following:

Team dynamics
: Demographics, conflict resolution, goal setting, psychological safety

Personality traits
: Extraversion, conscientiousness

Skill sets
: Programming skills, client management

---

1 They called it Project Aristotle as a tribute to the Greek philosopher, Aristotle, who is often quoted as saying, "The whole is greater than the sum of its parts."

Researchers conducted interviews and reviewed existing survey data for 180 Google teams. They used this data to run 35 different statistical models and understand which of the many inputs collected impacted team effectiveness.

Project Aristotle identified five key dynamics that contribute to the success of software engineering teams (see Figure 1-1). These are listed next in the order of their importance:

Psychological safety

This was the most important factor identified by the researchers. It refers to the extent to which team members feel comfortable expressing their opinions and ideas without fear of retribution or criticism. Teams that have high levels of psychological safety tend to be more innovative and take more risks, which can lead to better outcomes. The researchers found that when teams feel safe, they:

- Are less likely to leave the company
- Are more likely to utilize the diverse ideas discussed by the team
- Bring in more revenue and beat their sales targets
- Tend to be rated highly on effectiveness by their leadership

Dependability

This refers to the extent to which team members can rely on each other to complete their work and meet deadlines. Teams in which individuals trust each other to be dependable are more likely to be efficient and effective in their work.

Structure and clarity

These are conditions under which team members clearly understand the project's goals and their own individual roles and responsibilities. Team members who clearly understand what is expected of them tend to be more productive and focused.

Meaning

This refers to the extent to which team members feel that their work is meaningful and has a purpose. Teams with a strong sense of purpose tend to be more motivated and engaged.

Impact

This refers to how team members believe their work is making a difference and impacting the organization or society. Teams with a strong sense of impact are more committed to their work and the project's success.

Figure 1-1. Google's Project Aristotle: The five dynamics of effective teams

While Project Aristotle's research was conducted within Google, the identified factors influencing team effectiveness could hold some relevance for teams in other contexts. By focusing on these five factors, software engineering teams can create an environment conducive to collaboration, innovation, and success. As I'll discuss in Chapter 4, a good manager can foster these dynamics in their teams.

The researchers also discovered that variables such as team composition (size and colocation) or individual attributes (extroverted nature, seniority, tenure, etc.) did not contribute significantly to team effectiveness at Google. While these variables did not significantly impact team effectiveness measurements at Google, that doesn't mean they're unimportant, as indicated in the following section.

OTHER RESEARCH

While Project Aristotle is perhaps the best-known study on effective software engineering teams, many other studies have explored factors such as team composition, communication, leadership, and work processes. Here are a few key findings from some of these studies:

Smaller teams are better.

Although Project Aristotle did not recognize team size as relevant to team effectiveness, other studies (*https://oreil.ly/X7ryv*) have shown that smaller teams work better. As a team gets bigger, the number of links that need to be managed among members increases exponentially. Managing these multiple communication channels can be complicated. Many researchers have identified smaller teams containing less than 10 members as more likely to achieve success than larger teams.

Diversity can be beneficial.

It is sometimes suggested that team diversity may lead to communication and coordination problems (*https://oreil.ly/-F6aa*). For example, a diverse team would usually consist of people from different family backgrounds. Those with young children are more likely to seek flexible work hours, leading to coordination challenges. However, others have found that diverse teams can be more innovative and effective. A study by Lu Hong and Scott Page of the University of Michigan (*https://oreil.ly/-w5or*) found that groups of randomly selected (likely diverse) high-ability problem solvers can outperform groups comprising the best problem solvers. However, it's important to note that diversity alone is not enough. Teams must also create an inclusive and respectful environment for all team members. For example, a team that is supportive of team members who need flexible work arrangements will be able to coordinate better than a team that is intolerant of members with such needs.

Clear communication is vital.

Effective communication is essential for effective teamwork. Studies (*https://oreil.ly/JaG1H*) have found that teams that communicate frequently and openly are more successful than those that do not. The idea of psychological safety is a shared belief among team members that they can freely express their thoughts, ideas, concerns, or even mistakes without fear of negative consequences or judgment. Its importance is backed up by the research from Project Aristotle. Clear communication also provides the glue to connect team members and establish structure and clarity within the team.

Leadership matters.

The leadership of a software engineering team can have a significant impact on its success. Google's Project Oxygen showed that although teams could function without a leader, there is still a need for managers. It identified the essential traits that make for good managers and effective teams. I will talk about these traits in Chapter 4, but for now, it's necessary to understand that there is a strong correlation between effective leadership and positive team outcomes.

Agility enables adaptability.
Agility is the ability to adapt quickly to changing circumstances. In software engineering, this means being able to pivot when requirements change or when unexpected issues arise. Agile teams are quick to adapt and can work swiftly and efficiently while maintaining high quality. A study by McKinsey & Company (*https://oreil.ly/gxVuB*) found that organizations that underwent successful agile transformations reported a significant improvement in efficiency, speed, customer satisfaction, innovation, and employee engagement, all of which are essential to effectiveness.

Colocation powers innovation.
The debate over whether colocation or remote work is better for software team effectiveness is ongoing, with both approaches having their own advantages and disadvantages. Multiple studies conducted at Harvard (*https://oreil.ly/Tboto*), Stanford (*https://oreil.ly/Mw1fQ*), and others discuss the benefits of remote or hybrid work in terms of employee satisfaction and retention. However, other studies (*https://oreil.ly/4Qb8O*) have shown that face-to-face interactions at the workplace, both planned and serendipitous, trigger the flow of knowledge, sharing of values, and exchange of ideas (*https://oreil.ly/3JQSW*), which contribute to innovation.

While there may be trivial differences in the findings, we can build a theoretical picture of an ideal effective team based on the research findings discussed in this section. See Figure 1-2. By enabling psychological safety, clarity of structure and communication, dependability, meaningful work, and agility, software engineering teams can create an environment conducive to collaboration, innovation, and success.

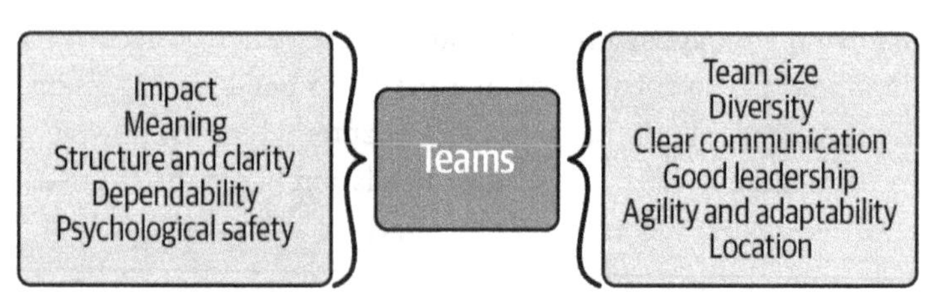

Figure 1-2. Factors that affect teams

You can now build on this understanding of dynamics and factors that influence the effectiveness of teams. The next things to consider are how the working environment can affect teams and how motivation can prime your team for success. As you go through the next sections, notice how the factors that affect teams pop up in various contexts.

Motivation Drives Performance

Before you can build an effective team or plan how to help make your existing team more effective, you need to understand and employ the power of motivation. By *motivation*, I'm not only talking about traditional rewards and incentives, such as compensation and tangible workplace perks. Such incentives can effectively motivate people to complete simple tasks. In contrast, *intrinsic rewards*, such as taking pride in your work and learning a new skill, are valued in modern workplaces where innovation and creativity are key—such as in software development. Intrinsic rewards motivate people to work on projects they're already passionate about, which could include both current projects as well as new, innovative ones. This validation and support allows these people to thrive and do their best work.

According to Daniel H. Pink in his book *Drive* (Riverhead Books, 2011), there are three elements that genuinely motivate people and drive performance:

Autonomy
: Autonomy is the desire to be self-directed and own one's work. Software engineering teams with a high level of autonomy tend to be more engaged and motivated, as each team can work in a way that best suits its individual strengths and preferences.

Mastery
: Mastery is the desire to improve one's skills and craftsmanship continuously. This principle is essential for software engineering teams as technol-

ogy constantly evolves and improves. Engineers committed to mastering their craft are more likely to produce high-quality work and contribute to their team's success.

Purpose

Purpose is the desire to do something meaningful and vital. This principle is essential for software engineering teams, as engineers often work on projects that significantly impact their business or industry. Notice that this echoes one of Project Aristotle's dynamics of effective teams: impact.

To build an effective team, you must consider these three catalysts that will help to motivate team members.

To understand this better, let me tell you a fictional story of a disillusioned engineer who was able to rediscover his passion and his empathetic manager who was able to motivate him. I shall call them David and Sarah.

David, once a bright-eyed software engineer at a tech company, was slowly losing his spark. He'd been there for five years, working on different projects, but the projects he worked on felt increasingly mundane. He missed the thrill of building something meaningful, something that wasn't just another cog in the corporate machine.

His manager, Sarah, noticed the shift in his energy. David wasn't the enthusiastic problem solver he used to be. During their regular one-on-one, Sarah gently inquired about his motivation. "It feels like you've lost your fire, David," she said. "Is there anything we can do to help you rediscover your passion?"

David hesitated. He wasn't sure how to express his growing dissatisfaction. But Sarah's genuine concern prompted him to open up. He spoke about his longing for meaningful work, his desire to create something impactful beyond another profit-driven application.

Sarah listened intently, nodding her understanding. She knew David was a valuable asset, and she wanted to help him reignite his passion. She suggested exploring options within the company that aligned with his interests, perhaps volunteering his expertise for internal projects with social impact goals. With Sarah's support, David began exploring different opportunities. He finally joined a team developing software for a renewable energy project, and his passion came flooding back.

He worked tirelessly, not for recognition or bonuses, but for the intrinsic reward of contributing to a cause he believed in. He was motivated by *purpose* and the firsthand impact of his work, knowing that his code was helping to create a cleaner future.

Sarah demonstrated good leadership skills in this case. She understood how motivation drives performance and used it to strengthen her team. She demonstrated empathy and collaborated with her employees and her fellow leaders to create a win-win situation for all.

As you can see, when someone is motivated, they are inherently driven to deliver high-quality results and are more likely to be effective. When building an effective team, identify how you can enable and activate the three elements that motivate and drive performance—autonomy, mastery, and purpose—into every step of the team-building process. For example:

- Let team members lead the development of individual modules so that they can develop autonomy and a sense of ownership.
- Help team members get access to the latest tools that will help them productively master the technology they are working on.
- Articulate how their contributions directly impact organizational goals so that they have a meaningful purpose.

Building an Effective Team

As discussed, effective teams share certain qualities or dynamics that enable them to be effective. Their performance is also driven by motivation. Let's now see how we can leverage this knowledge in practice to build these factors and motivations into an effective team. Regardless of whether you are working with an existing team, hiring new team members, or doing a mix of both, effective team building would usually require that you do the following:

1. Assemble the right people.
2. Enable a sense of team spirit.
3. Lead effectively.
4. Sustain effectiveness (a growth culture).

The steps outlined in this section are informed by the research findings on the key drivers of team effectiveness discussed earlier in the chapter.

Let's explore each of these steps in detail.

ASSEMBLE THE RIGHT PEOPLE

A modern software engineering team needs to be composed in a way that enables it to be effective. This means having the correct number of people with relevant skills, people with shared engineering mindsets, and people with diverse backgrounds and experiences, who can work together effectively to accomplish their shared goal. As the research on team size and diversity showed, the composition of a software team can have a big impact on its effectiveness.

If you're working with an existing team on an ongoing project, assess your current team structure and collection of skills and backgrounds. You may want to adjust things after reading this section to improve team effectiveness. If you're assembling a new team for a project, aim to build in size, mindset, and diversity from the start. You'll also want to consider this when adding new team members to an existing project.

While ideal team size and composition will vary based on factors like industry, company size, and product complexity, research provides some general guidelines. One study (*https://oreil.ly/PXL4Z*) found that the optimal team size for software projects is between three and five members. However, teams working on highly complex projects in large enterprise organizations may need to be larger, while startups or teams working on smaller-scale projects may be successful with even fewer members.

Hiring and interviewing for effectiveness

When building a new team or adding members to an existing one, it's important to assess candidates not just for their technical skills but also for the key mindsets and qualities that contribute to team effectiveness. During interviews, ask questions that probe for evidence of the engineering mindsets discussed earlier, such as caring about users, being a good problem solver, and being open to learning and growth. Present candidates with scenarios that test their approach to challenges like prioritization and balancing trade-offs.

To build a diverse team, cast a wide net in your recruiting efforts and be aware of potential biases in your hiring process. Set diversity goals and regularly review your pipeline and hiring decisions to identify areas for improvement. Throughout the hiring process, communicate your team's culture and values so candidates can self-select for fit. The hiring process is a two-way street, and you want team members who have bought into your team's ways of working.

Size

The composition of a software development team may vary depending on the complexity of the project and the development methodology being followed. In addition to software engineers or developers, there could be project managers, product managers, quality engineers, technical architects, team leaders, UI/UX specialists, etc. Each member has a responsibility to fulfill, and modern software engineering teams can vary in size from small 2-person teams to large teams with over 10 people.

Determining the right size for your team for every project is essential. The same set of people who worked very well together on a large project may falter on a small project. Consider this fictional story about a startup—I shall call them Code Crusaders. They had delivered a strong initial version of their product with a can-do spirit. But their initial success soon dissolved as their team size ballooned beyond manageable levels.

With 30 developers working on four different versions of the same product, communication became a tangled web of meetings, emails, and conflicting priorities. Deadlines were missed, projects stagnated, and frustration grew. The once-cohesive team fractured into isolated pockets, each working on its own piece of the puzzle without a clear vision of the whole. Decision making became agonizingly slow, bogged down by endless debates and conflicting opinions. The joy of collaboration was replaced by a feeling of being lost in a bureaucratic maze.

Despite individual talent, the Code Crusaders team found their effectiveness crippled by their growing size. They were a victim of their own success, a testament to the fact that bigger isn't always better.

The ideal size for your software engineering team depends on various factors, including the project, product, company culture, and team dynamics. Ask yourself the following questions:

Product

What resources do I need to build my product? Team size should reflect the resources required for a product. An app that needs to be updated regularly and has many users will require more resources than an internal tool that only one person uses.

Complexity

Is my product simple and easy to develop or intricate and tricky? A simple product, such as a chat client, will need fewer engineers compared to something more complex, like machine learning algorithms or AI systems.

Company culture
: How much autonomy does each member have within their role or team? Some organizations prefer several small groups and encourage collaboration among them. Others prefer a small number of larger groups working independently.

Leadership style
: Does the company encourage open communication between everyone involved so they know how their contributions fit into larger goals? Or does it focus more heavily on top-down decision making, where leaders make all major decisions without input from those below them who are closest geographically/organizationally?

Once you know the answers to these questions, you can check them against what you currently have or what you would want for your new team. You can then decide if:

- You need any additional engineers or engineers with different skill sets.
- The existing engineers on your team need to reskill.
- Outsourcing is required for some modules, or if they can all be developed in-house.

Shared engineer's mindset

To have an effective software engineering team, it would be ideal if each member had the right mindset. When I talk about *mindset*, I'm referring to the attitude of valuing the five dynamics identified by Project Aristotle (psychological safety, dependability, structure and clarity, meaning, and impact) and motivation for intrinsic rewards. This mindset is the glue that binds a group of software engineers together and makes them an effective team.

Software engineers have a clear high-level purpose: to build software that solves problems that customers would pay to have solved. Engineers must think about what matters most and the impact their software will have. Often, this involves delivering the best value to users and the business, and it also involves self-growth within the available time. As part of a team, engineers work in synergy to build software. Having a shared engineering mindset will ease the path toward synergy and effectiveness.

That said, in this section, I'll talk about qualities you should look for in software engineers when building or enhancing an effective team. Keep in mind that not everyone will feel the same way to the same extent about everything or excel in all of these attributes, and that's OK. The point is to have a healthy culture with characteristics of effective teams. However, by cultivating a diverse team, you may achieve a harmonious equilibrium, such that engineers can observe and learn from their colleagues within the team.

As such, when building a new team, hire engineers who already demonstrate most of the following characteristics. If you're enhancing an existing team, encourage and help your team to further develop these characteristics.

Cares about the user. When building a new team or building up your current team, you'll want engineers who care about users. An effective software engineer must understand that the user's needs are more important than using a specific technology or framework. A good software engineer will consider the following:

The problem domain
: What does the user want to accomplish with the product?

The business context
: What business purpose does the product support?

The business priorities
: What product features are more important for the business over the relevant timeframe (quarter or year)?

The technology
: What technologies are available, and which of these would work best for the product?

By asking these questions, a good software engineer will be able to contribute to a high-quality product that will fulfill the user's needs. If they don't think about how people will use the product or service, then it's likely that what's getting built won't be helpful in people's lives.

While this may sound obvious, I have seen developers who get caught up in the details of writing code without considering why they're doing so in the first place. This can lead them astray from their actual goals. Focusing on small tasks instead of contributing to significant outcomes like solving real problems or creating meaningful user experiences can prevent engineers from doing their best work.

If you're enhancing an existing team, develop this mindset in your team members by encouraging the following:

User interaction

While direct user interaction for each engineer may not be feasible, you can allow engineers to shadow support engineers or participate in usability testing drills to get acquainted with some day-to-day issues that users may encounter.

User research workshops

Encourage engineers to participate in workshops that focus on activities like user interviews or journey mapping. This can help them develop empathy toward user needs and perspectives.

Is a good problem solver. In a 2017 commencement speech (*https://oreil.ly/ROlUJ*), Dr. Neil deGrasse Tyson said, "You realize when you know how to think, it empowers you far beyond those who know only what to think." You can apply this principle to your engineering team because building software is not just about following established processes or knowing what to think; it's about fostering a mindset that encourages thinking beyond the ordinary.

In the same speech, Dr. Tyson shared an interview scenario where two candidates are asked about the height of a building's spire. The first candidate quickly provides the correct answer based on memorized information. The second candidate, however, admits not knowing the answer initially but demonstrates resourcefulness by measuring shadows to make an educated estimation. The second candidate in this case demonstrates the ability to think creatively and adapt to the situation to solve a problem.

Effective engineers have to be good at solving problems practically. Often, the best solutions are simple and elegant. But it may require thinking outside the box. Usually, a problem can be solved in multiple ways—some that are neat (but not overly complicated) and others that are unconventional but still effective.

The ability to solve problems also includes the ability to use prescribed tools and processes. An effective engineer should be able to solve problems within the constraints of the current technology while thinking out of the box if required. They should also consider all aspects of a problem at the same time.

Can keep things simple but cares about quality. Some engineers write overly complex solutions for use cases that may not necessarily exist. Although they want to be thorough, the low likelihood of these scenarios occurring may not justify the effort spent. Effective engineers should understand the core problem and know how to solve it reasonably while keeping things simple. It's also necessary for them to understand the trade-offs between simplicity and performance.

Although this trait may sound deceptively simple, it requires constant vigilance, as it may be necessary to balance trade-offs between different quality dimensions (e.g., accessibility versus performance). When competing priorities are in play, engineers should be able to make informed decisions based on their knowledge of the product domain and its users.

Encourage engineers to clearly define the core problem and specific use cases before diving into solution development. Doing this will help them avoid overengineering and deliver simple yet high-quality solutions.

Can build trust over time. Effective engineers should understand the importance of being trustworthy, as in they can be depended upon to do their tasks as expected. Trust cannot be built overnight, so demonstrating trustworthiness for a period of time also shows dependability and consistency. This understanding of trust leads to engineers having autonomy and social capital:

Autonomy
: Autonomy is built on trust. Someone is given autonomy if they can be trusted to do their tasks dependably and consistently. As mentioned earlier, teams with a high level of autonomy tend to be more engaged and motivated. Autonomous workers are happy workers—and happy workers tend to be more productive than their coworkers who aren't given any choice in how they do their job. But autonomy doesn't mean they can misuse their resources or power. Self-motivated engineers who can study a given situation to decide on the best way to address it are more effectively autonomous. Autonomy also does not translate to asking fewer questions. Committed individuals would promptly seek clarifications to unblock themselves and the task at hand in effective ways.

Social capital
: You have *social capital* when you have a network of positive relationships with people. These relationships are built on cooperation and trust. Effective software engineers build networks of positive relationships with people. These engineers are adept at sharing their skills with others within and outside their team and enabling collaboration.

An effective software engineer understands that trust is built over time through mutual respect and open communication with other team members. They can make good with any autonomy granted to them while still collaborating effectively on projects requiring creativity and technical skill—and maybe even some good old-fashioned fun!

You can help your existing team members to develop these skills by encouraging them to participate in initiatives that span beyond their core roles and not only showcase their expertise but also foster relationships and trust with team members from diverse backgrounds. At Chrome, I frequently encourage my team members to write about the new features they are developing on the Chrome developers blog (*https://developer.chrome.com/blog*) or engage with the developer community through tech talks. This collaborative experience not only builds their social capital but also grants them a sense of autonomy as they take ownership of their contributions.

Understand team strategy. Effective engineers should understand and be able to communicate how the team will achieve its goals and how their actions can help or hinder its success. They should be able to answer these questions:

- What are we trying to accomplish?
- How do we plan on getting there?
- What role do I play in achieving these goals?
- How will my actions affect other teams or individuals in attaining their goals?

Whether you are interviewing someone new or conducting a one-on-one session with an existing team member, consider asking them questions such as "How do you see your skills contributing to our objectives?" or "What potential challenges do you foresee?"

Imagine you are managing a software development team tasked with reducing application load times to enhance user experience. As a manager, you could outline the strategy during a team meeting, emphasizing that faster load times are crucial for retaining and attracting users. You might also discuss the specific roles team members will play, such as optimizing code, improving server performance, or conducting user experience testing. This involvement helps team members connect their daily tasks to the broader goal of enhancing user experience. They are more likely to understand the team's strategy and feel like they are a part of it, which in turn enhances their sense of purpose.

Can prioritize appropriately and execute independently. Effective software engineers can prioritize appropriately and execute independently. Software engineers are often required to juggle multiple priorities, including technical debt, speed of delivery, and quality, while keeping the business goals and ultimately the customer in mind. When faced with issues, effective engineers don't rush into quick fixes. Instead, they seek the root cause of problems and propose solutions that balance these priorities effectively.

A competent software engineer also knows when it's appropriate to take ownership of tasks and projects without being told what needs to be done by their direct manager or leader. They will know when drawing from their peers for help bearing this load makes sense—and communicate effectively with other teams when necessary.

You can help team members prioritize their work by clearly sharing the organization's and project's strategic priorities and following up with them through regular check-ins and goal-setting sessions. Always ensure that engineers have the necessary resources and guidance to execute their responsibilities effectively.

Can think long term. Thinking long term is an essential trait for an effective software engineer. They need to see the big picture and understand how each product component fits into the company's overall goal or set of goals.

Software engineers who can think long term are better equipped to understand how new technologies will impact their products in the future. They also better understand what kind of team dynamics and culture will affect their development efforts. They design and build flexible solutions, anticipating the changes that may be required in the future.

When enhancing an existing team, you should positively acknowledge the actions of forward-thinking team members and encourage other members to follow suit. A simple example could be variable naming conventions in code. Using variable names such as *x* and *y* may be quicker but reduces the maintainability of the code for someone looking at it six months down the line. Encourage your team to write code that can be understood by other developers too.

Can leave software projects in better shape (if time allows). If time allows, engineers who are maintaining existing code or projects should be willing to leave a project in better shape than they found it. This means making the code better, making sure that the documentation is up to date and accurate, cleaning up the environment so it's easier for the next person to pick up, improving the team's processes and culture over time, and growing as a professional developer by learning more about other languages/technologies/frameworks that are relevant.

They should also consider how their work impacts their immediate colleagues and everyone else in the organization and community. Identify individuals in your team who suggest cleaning up the surrounding code when working on an existing module. Ask them to review code written by other engineers so that they can mentor the others. When hiring a new team member, ask them to share examples of times when they improved the quality of a project that they worked on.

Is comfortable taking on new challenges. An effective software engineer is comfortable taking on new challenges if organization or team requirements change. This could mean taking on a new project or additional responsibilities such as mentoring or reviews within the same project.

As technologies evolve, so do the requirements for the job. Flexibility and the ability to learn quickly will ensure that engineers can adapt when necessary and grow with the organization.

When enhancing an existing team, keep an eye out for team members who have been working on the same part of the code for a long time and assign them fresh tasks or responsibilities that encourage growth. This approach not only fosters growth and engagement but also helps engineers shed any fear of change, enabling them to comfortably embrace new challenges.

When hiring members for a new team, consider individuals who possess some of the required technical skills but also demonstrate openness to exploring new technologies or domains.

Can communicate effectively. Communicating effectively within the team, with other teams, and with the customer is essential. This means that engineers must be able to communicate technical information effectively to peers and management. This quality is important because miscommunication may result in bugs or issues that need fixing.

When communicating with peers, engineers should use clear and concise language. They should also pay close attention to what their peers say to help them understand their colleagues' perspectives and concerns.

Communicating effectively with other stakeholders, such as sales or marketing departments, is also essential. A software engineer should know what their product does, how it works, and why it benefits customers and users.

When communicating with customers, engineers should use simple language so the customers can understand them clearly without getting confused by technical jargon. Otherwise, customers will lose interest very quickly!

To improve communication in an existing team, encourage team members to share ideas freely, raise concerns, ask questions, listen actively, engage in constructive discussions, and value diverse perspectives. Also consider organizing workshops for those who need additional help with their communication skills.

If you are interviewing new engineers, consider candidates who can articulate technical concepts clearly and concisely. Evaluate their ability to listen actively and respond thoughtfully. Candidates who can adapt their communication style to different audiences and demonstrate an understanding of cultural nuances are able to adapt better in existing teams.

Diversity and inclusion

When building a new software engineering team or enhancing an existing one, aiming to have team members with diverse skill sets, backgrounds, and experiences makes a difference. This can include diversity in terms of race, gender, age, educational background, and work experience. For example, if you have a group of engineers who all come from an enterprise background, consider rounding out your team by finding someone who has experience building consumer apps as well.

By developing a diverse team, you'll be able to tap into a broader range of perspectives and experiences, which can lead to better problem-solving and innovation.

Google Translate (*https://research.google/teams/language*) and other natural language processing products are testaments to how diversity in a software team results in a successful product. Computer scientists and linguists have worked hand-in-hand on the Google Translate team. The linguists on the team provided insights into the nuances of different languages and how they could be translated effectively. At the same time, computer scientists and engineers were able to develop the algorithms and technology needed to power the translation engine.

Having a diverse team alone is not enough. Building diverse teams requires intentional effort, overcoming biases, and navigating cultural differences. There is bound to be some friction, and enabling seamless collaboration among a group of individuals with distinct personalities is challenging. Yet, with patience, understanding, and a commitment to inclusivity, the initial friction can be overcome to build a strong cohesive unit. I have been fortunate enough to experience this firsthand.

It would help if you also created a culture of inclusion.

For example, at Google, we initiated a new team for building a new developer product. The diversity was palpable. We had members from four continents, ranging from fresh-out-of-college juniors to veterans with over a decade of experience.

However, the initial meetings were more like a monologue than a dialogue. The junior members, smart yet unseasoned, were reluctant to voice their ideas, often overshadowed by more experienced voices. I used to regularly hear comments like "I'm not sure that what I have to add is going to matter compared to what the senior engineers have to say." I had to plan to overcome these challenges.

My first step was to create an environment where every opinion was valued equally. I initiated round-robin sessions, where each member, irrespective of their rank or tenure, was given uninterrupted time to voice their thoughts. This not only encouraged junior members to speak up but also helped senior members to listen actively.

Second, acknowledging and embracing our cultural differences played a pivotal role. I started having biweekly virtual "cultural exchange" meetups. Team members would share something unique about their culture—be it a local tradition, a festival, or even a coding practice unique to their region. This not only broke the ice but created a tapestry of cultural awareness and mutual respect.

Third, I promoted psychological safety through open forums. Psychological safety doesn't sprout overnight. It's a garden that needs constant nurturing. To this end, I established an "Ideas and Concerns" forum. Here, team members could anonymously post their ideas or concerns. Every week, we would address these in our team meetings, ensuring that each voice, however quiet, was heard and considered. This practice encouraged even the most introverted members to share their innovative ideas without the fear of judgment.

Finally, to further bridge the experience gap, I paired junior members with senior mentors. These mentorship relationships went beyond mere technical guidance. They included navigating workplace dynamics, understanding unwritten industry norms, and developing soft skills crucial for career progression.

These strategies transformed our team dynamics. The once hesitant junior members began contributing ideas that were out of the box and sometimes crucial to solving complex problems. The senior members, on the other hand, found fresh perspectives and renewed enthusiasm in mentoring. Our project not only met its deadlines, but subsequent releases went even more smoothly.

Through this journey, the most significant lesson I learned was that leadership in tech is not just about managing projects; it's about nurturing an ecosystem where diverse talents can coalesce, grow, and create something extraordinary. This means creating an environment where all team members feel valued and respected, regardless of their background or identity. Some strategies for creating a culture of inclusion include:

- Providing diversity and inclusion training for all team members to help them understand different backgrounds and practices
- Encouraging open communication and feedback
- Ensuring that all team members have equal opportunities for growth and development
- Providing flexible work arrangements to accommodate diverse needs and lifestyles

To wrap it up, an effective software engineering team needs three important things: the right number of people, team members who have the right attitude for being effective both by themselves and as a team, and people with different skills and backgrounds. These factors together help you create a foundation for effective teamwork and success.

ENABLE A SENSE OF TEAM SPIRIT

Once your team has been assembled, the next step to building an effective team is enabling a sense of team spirit. A group of people has *team spirit* when they feel invested in reaching their shared goal and are there to support each other. It is more than just getting along or being helpful. As previously stated, by definition, a team is bound by shared responsibilities and goals. Members work together and share mutual accountability to solve problems and achieve common goals. With team spirit, members work together in harmony but are also willing to help each other accomplish tasks. A group that works in harmony can get more results in less time than one that does not.

Cultivating team spirit is all about building the key effectiveness factors of psychological safety and dependability that were identified in the Project Aristotle study.

You can cultivate team spirit by creating an environment for collaboration and communication. The foundation of this involves defining roles and responsibilities, establishing a shared purpose, and fostering trust among members.

Define roles and responsibilities

Defining roles and responsibilities within a team is a fundamental step toward effective collaboration. It isn't just about assigning tasks; it's about ensuring that each team member understands their specific duties, which in turn minimizes confusion, the risk of overlapping efforts, and undue frustration.

Software roles do have some overlapping responsibilities depending on context. For example, establishing responsibilities for different types of documentation such as requirement specifications, test cases, user guides, API documentation, etc., can be challenging.

My friend Alex was once assigned to lead a project where developers were efficient at churning out code but avoided updating design documents. Similarly, testers often found themselves struggling to decipher requirements generated by a collaboration between the business analysts and technical writers. Meanwhile, the technical writers themselves were swamped with internal tasks and did not have enough time for the user manual. The project was on fire and, in spite of having fully functional code, users were unhappy.

Alex knew the problem: misaligned responsibilities. He implemented ownership. Developers, empowered with clear expectations, penned their own docs and unit test cases. Business analysts worked with product owners to define clear requirements. Testers, equipped with solid requirements, focused on preparing clear test plans and finding issues. Technical writers, freed from internal burdens, turned their attention to crafting user-friendly guides.

The transformation was subtle but impactful. Collaboration replaced blame. The once-dreaded testing phase was now embraced as an opportunity to collaborate and improve the product. User manuals turned out to be informative and intuitive. Alex had successfully demonstrated the importance of ownership and role clarity.

Role clarity can thus help to cultivate a sense of team spirit and unity if you align responsibilities with individual strengths and skills. You must also highlight the interdependence of team members' contributions, emphasizing the collaborative aspect of their work. This approach not only enhances efficiency but also fosters a culture of mutual support and shared achievement.

Moreover, role definition should allow for skill development and adaptation as the project progresses. It's essential to recognize and appreciate the contributions of each team member, reinforcing their sense of value within the team. By periodically reviewing and adapting roles and responsibilities, you can ensure that they stay aligned with evolving needs and individual growth. In this way, defining roles and responsibilities becomes a dynamic process that not only promotes clarity but also nurtures team spirit and maximizes effectiveness.

Establish a shared purpose

Similar to having a shared goal, having a shared purpose drives the team to work together and overcome their differences. While goals help to convey what the team should be achieving, purpose articulates why they need to achieve these goals. I have seen many software engineering teams struggle to balance competing priorities, with architects focused on scalability and performance, developers on deadlines and code efficiency, testers on identifying edge case issues, and designers on aesthetics and experience. While each of these is important, teams need a shared sense of purpose to align on what to prioritize to create the best overall product. While focusing on all of these is essential to create a good product, a shared purpose is required to understand the priorities.

Without a shared purpose that balances ambition, quality, user experience, and design, these teams face challenges such as technical debt, scope creep, usability issues, and internal conflicts.

A leader who establishes a shared purpose is able to channel all the different energies to create a platform that's both innovative and user-friendly, delivered with quality and efficiency.

To establish a shared sense of purpose, you can do the following:

Communicate the overall project purpose and goals
: Clearly articulate the project's purpose, target audience, and desired impact. Share user personas, mock-ups, or competitor analyses to create a tangible picture of the "what" and "why." This provides a North Star for individual contributions.

Encourage team members to share their ideas and feedback
: Organize brainstorming sessions, workshops, and regular check-ins where developers, testers, designers, and architects can freely share their expertise and concerns. This fosters a collaborative environment and surfaces diverse perspectives, enriching the shared vision.

Ensure that all team members understand how their work fits in with what their teammates are doing and into the larger project goals
: Link individual tasks to specific features and user stories, and highlight how each team member's contribution impacts the product's overall functionality and user experience. This creates a sense of shared ownership and reinforces the "we're in this together" mentality.

You can achieve these objectives via regular knowledge-sharing sessions or team-bonding exercises within or outside the office. These activities create opportunities for team members to come together, exchange ideas, and build rapport beyond their immediate tasks. It fosters a sense of belonging and reinforces the shared purpose, reminding everyone that they are part of a collaborative effort aiming for a common goal. As a bonus, these sessions can also ignite fresh perspectives and innovative solutions. By taking these steps, you will increase the chances of successfully delivering a product that meets user needs while achieving the team's ambitions.

Foster trust among team members

An essential part of cultivating a sense of team spirit is building trust among team members. To understand this better, imagine there are two teams; let's call them Team Open-Door and Team Silos.

Team Open-Door is known for its collaborative spirit and open communication. Shared knowledge and a culture of asking for help without judgment leads to faster problem-solving in this team. Trusting fellow teammates to do the right thing reduces stress levels and helps to boost morale. Open discussions lead to diverse perspectives, leading in turn to innovative approaches and unexpected features. It is a team that attracts top talent and accolades from clients and is trusted by management to handle key projects.

Team Silos, on the other hand, consists of highly skilled but isolated developers, each working on their own modules with minimal communication. Miscommunication and rework due to conflicting code and overlapping functionalities slow them down frequently. A "not my problem" attitude further delays the resolution of issues. Frustration and resentment fester in a blame game environment. The lack of trust hampers collaboration and leads to subpar results.

Overall, trust within a software engineering team isn't just a feel-good factor; it's a powerful driver of success. Open communication, shared knowledge, and a culture of collaboration can turn a team of individuals into a cohesive unit, producing outstanding results. Trust can also help to foster better work-life balance for individuals on the team. You are able to take short breaks from work and feel relaxed during these, when you trust other team members to fill in for you.

To foster trust within your team, you can:

- Encourage open communication and feedback.
- Provide opportunities for team members to get to know each other on a personal level.

- Be transparent about project goals and timelines.
- Reward teamwork and collaboration.

Working together toward a shared goal with fellow team members you trust to have your back goes a long way toward binding the team, enabling a sense of team spirit, and motivating it to stay on track. While such a team may seem auto-motivated, a leader is still needed to help clear roadblocks and provide mentorship and guidance. Let's see what it takes to lead effectively.

LEAD EFFECTIVELY

To build an effective team, you must also be an effective leader. Although strong teams can function without a leader, effective leaders impact employee performance and satisfaction, decision-making skills, and collaboration, and they promote a positive work environment.[2]

The Project Oxygen research highlighted the crucial role that leadership plays in team effectiveness. You need to fulfill your core responsibilities, enable effective practices, and use strategic visibility. You'll explore what it means to be an effective leader and enable effectiveness in the coming chapters of this book, but this section will touch upon the core responsibilities of a leader and the importance of strategic visibility.

Responsibilities of effective leaders

As a leader, your core responsibilities are to inspire, influence, and guide your team toward a shared goal. Effective leaders build successful teams by incorporating effectiveness practices into these core responsibilities. For example:

- You are responsible for planning team roles and composition.
- You help to set goals and priorities for the group/team members.
- You ensure that everyone has what they need (tools/resources) to complete tasks effectively.
- You leverage your experience to provide insights into where issues may arise so they can be proactively addressed before they cause significant problems.
- You establish clear lines of communication using tools and processes that promote effective communication.

---

2 I'll talk more in detail about the effect of manager behaviors in Chapter 4.

- You communicate regularly via team meetings, status updates, and progress reports. Communication should be timely, transparent, and inclusive.

Notice that these responsibilities directly support the characteristics that are built into an effective team.

Strategic visibility

Strategic visibility is about communicating the team's accomplishments and their impact on the business to internal and external stakeholders. This can involve highlighting how their work addresses a critical customer need, improves efficiency, or contributes to a larger product vision. By effectively showcasing their value through compelling stories and data, you ensure that the team gains recognition within the organization and is positioned for future opportunities, resources, and influence.

As previously discussed, motivation drives performance. Although people can be motivated by autonomy, mastery, and purpose, they are also motivated by recognition and validation. As a leader, make a point to publicly support the efforts of your team within your organization—without calling attention to your own involvement. Not only do organization-wide recognition and acknowledged success fuel team motivation, but they also showcase talent and unlock doors to future growth opportunities for your team members. Remember, you can be your team's best cheerleader, inspiring them to reach new heights and achieve their goals together.

Here's a real-life example to better illustrate strategic visibility. I once led a team of exceptionally skilled engineers. Despite our consistent delivery of high-impact work, we were like a hidden gem, often overshadowed by the more glamorous, high-profile projects within the company. This was the case even though (we thought) our contributions were integral to the overall product's success.

Initially, I observed a tinge of frustration among my team members. They worked tirelessly, often solving complex problems, but they did not receive the recognition they deserved. As a leader, I knew it was crucial not only to acknowledge their work but to ensure it was seen and appreciated at a broader organizational level.

I realized that the key was not just working hard but working smart—aligning our efforts with the company's top priorities. This didn't mean abandoning our current projects but rather finding a way to connect our work to the larger narrative of the company's goals.

We started by identifying a project that was directly tied to one of the company's main objectives for the year. This project wasn't just a high priority for the company; it was also a perfect match for our team's unique skills and expertise.

We dove into the project, applying our deep knowledge of browser technologies and product engineering principles. We focused on not just completing the project but exceeding expectations, adding innovative features that we knew would make a significant impact.

I encouraged the team to document our process and outcomes meticulously. We shared regular updates not just within our team but with other teams and stakeholders. This transparency wasn't just about visibility; it was about opening channels for collaboration and feedback.

The project was a success, significantly impacting the company's goals. But more importantly, it put our team in the spotlight. Our work was recognized in company-wide meetings, and team members were invited to speak at internal tech talks and conferences. This recognition was a morale booster, and it fostered a sense of pride in our team.

Here's what I learned about leadership from this experience. As a leader, you should do the following:

Align with broader objectives
: Your work's impact multiplies when it aligns with your organization's primary goals.

Leverage your team's unique strengths
: Understand and utilize the unique skills and knowledge of your team.

Communicate effectively
: Regularly share your progress and learnings. Transparency fosters recognition and collaboration.

Focus on impact, not just hard work
: It's not just about working hard but also about creating tangible impacts that align with organizational priorities.

Build a narrative around your work
: Document and share your journey; it's crucial for visibility and recognition.

In a nutshell, this experience taught me that, in the tech industry, being in the shadows doesn't mean you're not valuable. Sometimes, it just means you need a strategic approach to showcasing your value. By actively championing your team members, you cultivate a culture of trust, loyalty, and mutual support, ultimately empowering them to excel and contribute to the overall success of the team.

SUSTAIN EFFECTIVENESS (A GROWTH CULTURE)

The final step to building an effective team is sustaining the culture that you have built via continuous growth. Establishing a shared purpose and open communication can foster a strong team culture, leading to an engaged and motivated team. Additionally, for team members to believe that their impact and contributions matter to the organization, leaders must create opportunities for team members to develop their skills and grow within their roles. Sustaining effectiveness over time (as seen in Figure 1-3) requires supporting the factors of agility, purpose, and impact that researchers have found to be key drivers of team success.

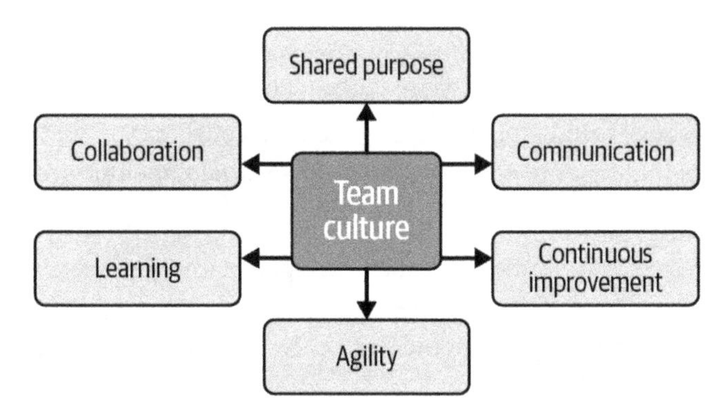

Figure 1-3. Sustain effectiveness: improve and grow

Learning and development opportunities

Growth opportunities can be great sources of motivation for team members, and, as discussed earlier in this chapter, motivation drives performance. On the other hand, lack of learning and development opportunities can lead to stagnation. A team that does not know how to improvise can get stuck in outdated methods. Imagine a team today that has not considered using AI tools to improve its processes. Stagnation can also breed boredom and frustration, leading to decreased productivity and talent drain.

Growth can also come in the form of learning opportunities, which can be created through training, mentorship, and coaching.

Agility

As discussed earlier in this chapter, research has shown that agility is critical for team effectiveness because it enables teams to respond quickly and effectively to changing business requirements, customer needs, and market dynamics. Agile teams are more adaptable to changing circumstances and can adjust their development plans when needed.

Here are some strategies for ensuring team agility:

Emphasize agile methodologies
: Agile methodologies provide a framework for delivering software flexibly and iteratively. By implementing agile practices, such as Scrum or Kanban, teams can improve their ability to adapt to changing requirements and priorities. However, it's essential that you tailor agile processes to meet your team needs rather than follow them blindly from a textbook.

Promote cross-functional collaboration
: Encourage team members to work across functional boundaries, such as developers working with UI/UX designers. This can help break down silos, encourage sharing and reuse, and increase collaboration and communication, leading to more efficient and effective development processes.

Prioritize communication
: Effective communication is critical for agile teams. Ensure team members have regular check-ins and meetings to discuss progress, challenges, and priorities. Encourage open and honest communication, and provide channels for feedback.

Build a culture of adaptability
: Foster a culture that values adaptability and embraces change. Encourage experimentation and risk-taking, and reward teams for being responsive to changing circumstances and customer needs.

Implement continuous integration and delivery
: Continuous integration and delivery practices can help teams deliver software faster and reliably. By automating the building, testing, and deployment process, teams can reduce the time and effort required to release new features and updates.

By implementing these strategies, effective teams can become more agile, adaptive, and responsive to changing business needs and customer requirements.

Continuous improvement

Effective software engineering teams are committed to continuous improvement. This means they constantly seek ways to improve their processes, tools, and skills. Teams should regularly evaluate their performance and identify areas for improvement. They should also seek feedback from other teams, stakeholders, and customers to identify areas for improvement. Encourage team members to be committed to continuous improvement and provide them with the tools and resources they need to be successful. This can include access to new technologies, feedback loops, and a culture of experimentation.

Continuous improvement requires a culture of learning and growth. The following tips can help to create such a culture:

Foster continuous learning

Encourage team members to learn new skills and technologies and provide them with opportunities for professional development. This can help team members stay up to date with industry trends and best practices.

Measure and monitor performance

Regularly track key performance indicators (KPIs) to assess team performance and identify areas for improvement. Use the data to inform decisions, prioritize work, and make adjustments to ensure the team meets its goals.

In an ever-changing technology landscape, software development teams must continuously adapt to meet evolving user needs, business requirements, and market demands. Agile methodologies can promote a culture of continuous learning and improvement within the team and enable software development teams to be more efficient and effective in delivering value to stakeholders.

Conclusion

Building an effective software engineering team takes work. As you've seen in this chapter, many factors can influence the success of a software engineering team.

The guidance provided in this chapter on building effective teams maps closely to the research findings on the key factors that drive team effectiveness.

Composing a strong team builds the foundation for the right mix of skills and perspectives. Enabling team spirit fosters the psychological safety and dependability that allow team members to take risks and rely on each other. Effective leadership provides the direction and support teams need to do their best work. And sustaining a growth culture allows teams to continuously improve and adapt to new challenges.

Project Aristotle and other research on effective teams showed that enabling psychological safety, clarity of structure and communication, dependability, meaningful work, and agility can create an environment conducive to collaboration, innovation, and success. These factors recur throughout the chapter and bind teams together in their drive to be effective.

Effective teams share certain qualities or dynamics that enable them to be effective. Their performance is also driven by motivation. Therefore, to build an a new team that will be effective or develop an existing team to make it more effective, you must take certain factors into consideration. These factors include composing your team with the right number of people for a project, with diverse skill sets and backgrounds. It also includes developing an engineering mindset in each member, which creates a synergy that eases the way for collaboration and effective performance. This is further cemented by a sense of team spirit.

Although it may sound obvious, an effective team must be led by an effective leader. Set up your team for success by incorporating effectiveness practices in everything you do, and don't be afraid of acknowledging your team's efforts. Recognition is valuable, not just for morale, but for your team's career growth.

Finally, to have an effective team, you must sustain a growth culture. This keeps the team agile, always improving, and better equipped to handle anything that comes its way.

These foundations can help us structure our teams for the best possible outcomes. These factors can help you create a motivated and productive team that consistently delivers high-quality results. With the right people, communication, and support, you can build a team that thrives and successfully completes even the most complex projects.

Measuring and monitoring effectiveness throughout a project's lifecycle is as essential as enabling it. To correctly measure effectiveness, you need to understand how it differs from concepts of productivity and efficiency. The following chapter discusses the differences between these concepts and how they are measured.

2

Efficiency Versus Effectiveness Versus Productivity

The previous chapter provided an introduction to what makes a software engineering team effective and how to build those characteristics into your team. By now, you must have an idea of what it means to be effective. *Effectiveness* is an indicator of individual or team performance. You may have also encountered the terms *efficiency* and *productivity* in relation to employee or team performance, but what do those terms actually mean? While these concepts are interconnected, they hold distinct meanings. Understanding these differences is crucial for accurately assessing and guiding your team.

A software engineering team that is churning out code is efficient if it's following processes, writing clean code, fixing issues, and meeting deadlines. It's also productive if it's quick in completing its tasks of coding, testing, fixing, and releasing the code. However, to be effective, the code released should address user problems and make a positive impact on the business. Simply put:

- Efficiency is about doing things right.
- Effectiveness is about doing the right thing.
- Productivity is a measure of output over input.

While you can use any of these concepts to measure a team's performance, it's essential to understand the implications of each. This chapter will dive into the differences between efficiency, effectiveness, and productivity for software engineering teams. You will explore the factors influencing each of these and methods to measure the impact of knowledge work. Finally, I will discuss the importance of focusing on outcomes rather than outputs and how you can shift your team culture to strike the right balance between effectiveness and efficiency.

The Differences Between Efficiency, Effectiveness, and Productivity

To better understand the differences between efficiency, effectiveness, and productivity, let's compare their purposes, how to measure them, and the factors that influence them.

GOALS

Let's start with detailed definitions of efficiency, effectiveness, and productivity that focus on what each term implies and how they differ in terms of goals:

Efficiency
: Efficiency implies *doing things right* to minimize waste and maximize output. Software engineering teams improve efficiency by adopting iterative methodologies, implementing continuous integration and delivery pipelines, and using tools such as code editors, debuggers, and performance profilers. Thus, to become efficient, a software engineering team could maximize the number of features it fully develops with the given resources. A good measure of a team's efficiency is the number of features it fully develops and tests per unit of time. For example, "We released three features this week; each one took us three days."

Effectiveness
: Effectiveness implies *doing the right thing* and delivering the right outcome. The right outcome provides value to the organization and its customers and is focused on their needs. Teams can achieve this by aligning their goals with the organization's objectives, defining a clear strategy, and focusing on outcomes. Software engineering teams can improve effectiveness by prioritizing features and initiatives that align with the organization's vision, using metrics to measure the impact of those features, and seeking feedback from customers and stakeholders. You can measure the effectiveness of software by tracking key metrics like user adoption rate and customer satisfaction scores, among others.

Productivity
: Productivity is a subset of efficiency and *measures output over input*. It is about executing at a pace that enables teams to achieve their goals quickly. Software engineering teams can improve their productivity by increasing

their throughput, velocity, and code output. However, it is essential to note that traditional definitions of productivity do not often consider the quality and impact of output. Productivity is often used to measure the efficiency of machines and capital. However, it cannot accurately estimate human knowledge work.

Table 2-1 gives a summary of how efficiency, effectiveness, and productivity differ from each other.

Table 2-1. Efficiency, effectiveness, and productivity goals

	Efficiency	Effectiveness	Productivity
Goals	Doing things right	Doing the right thing	Output over input
Measurement	Time Resource utilization Quality	Customer satisfaction Business value User adoption rate	Lines of code Delivery of function points Delivery of story points

To achieve efficiency, effectiveness, or productivity, an engineering team may need to target a different set of goals overall. They may mean different things, but it is important to note that these concepts are not mutually exclusive. To understand this better, let's consider this fictional story of the "Cloudoids," an engineering team that had recently been tasked with the migration of their company's flagship software to microservices and Kubernetes. The team had buzzed with excitement about this opportunity to learn a new technology while working on something that was important to the business.

The product manager published a backlog of features that were to be migrated, and the team attacked it with fervor. Led by Maia, who had a penchant for pushing boundaries, the team hit the ground running. Team members were enthusiastic about developing features, committing code, fixing issues, and deploying builds to the new architecture sprint after sprint. Each milestone was met with a flurry of high fives and pizza to celebrate the fact that they were meeting their deadlines while delivering bug-free code.

But amid the celebrations, user feedback started whispering doubts. No performance improvements, no usability leaps. Maia, a data-driven leader, unearthed the truth: metrics revealed they'd built a tech wonderland devoid of impact. While they had improved scalability and could release builds faster, they had not leveraged the technology for users' benefit. To the users, the product has remained more or less the same. The team had focused on "how" while neglecting the "why."

This triggered a shift to focus on the right outcomes. Team members retraced their steps to find how users could benefit from the change in architecture. Gradually, they improved their design to support efficient and intuitive use. Metrics proved that they were finally using the technology the right way to achieve the correct outcomes. The Cloudoids had learned: effectiveness wasn't just about speed or shiny tech. It was about weaving efficient technology with purpose.

The story teaches us that, by focusing on the development of the right definitions and metrics for efficiency and productivity in their specific context, teams can concurrently fuel their overall effectiveness.

MEASUREMENT

Although efficiency, effectiveness, and productivity all reflect performance, you must consider different things when measuring them. Both productivity and efficiency depend on the output of a specific activity. However, productivity is a raw measure, while efficiency is a refined measure that depends on the outcome. Effectiveness also depends on the outcome. I'll talk about the difference between output and outcome in a later section, but for now, let's focus on the traditional measures of efficiency, effectiveness, and productivity.

Productivity measures output over input. Traditional measures of team productivity in software engineering include the following:

Lines of code
: This involves counting the number of lines of code a developer writes, presuming that the more you write, the more you accomplish.

Function points
: This is based on the functionalities provided by the software. It considers the complexity and number of features the software offers.

Story points
: This measures the complexity of user stories in Agile software development. Story points are assigned to each user story based on its complexity and required effort.

DevOps metrics
: DevOps metrics such as lead time, deployment frequency, mean time to recovery, and change fail rate focus on the speed of the software development process.

Productive software engineering practices will result in more output in terms of the amount of code delivered by the team. Depending on the metric used, it can mean the volume of lines of code or function points delivered is more. Focusing on productivity alone can be problematic, as we will discuss later in the chapter.

Efficiency enhances the productivity measurement by including other contributing factors such as the following:

Time
: A process is more efficient if tasks are completed faster.

Resource utilization
: An efficient team better utilizes resources such as time, money, and personnel.

Bug fix rate
: Faster detection and fixing of bugs indicates an efficient team.

Defect density
: Defect density is the number of defects in the code per line of code. A lower defect density indicates higher efficiency.

Quality
: Efficient teams produce high-quality products.

Efficient software engineering practices will result in less wastage of resources such as time and money. Efficient teams are productive and generate the desired output by optimizing their resource usage.

Effectiveness measures the ultimate outcome of the development activity. Some standard measures of team effectiveness in software engineering are as follows:

Customer satisfaction
: This is the degree to which the software product/service built by the team meets the needs and expectations of its users.

Business value
: This is the contribution of the software to the overall goals and objectives of the organization.

User adoption rate
: This is the extent to which the software is used by its intended users.

ROI (return on investment)

This is based on the net profit earned versus the total cost of developing the software.

Time to market

This measures the efficacy of the software development and release process.

Effective software engineering practices will result in higher customer satisfaction, business value (*https://oreil.ly/7RRCe*), user adoption rate, and ROI. Similarly, a shorter time to market will also add to the product's value, making it more effective.

To understand this better, consider the following scenarios where two teams, A and B, are developing an MVP (minimum viable product) for a food delivery app:

Team A has 10 engineers who delivered the initial MVP for a new app in 30 days. Team B has 8 engineers who delivered similar functionality in 40 days. Which team was more productive?

Team A is more productive because it generates more output over the given time.

(Assuming an 8-hour workday, Team A worked 2,400 hours and Team B worked 2,560 hours. Since the delivered functionality is similar, Team A is more productive.)

Team A's MVP had five major usability issues, which took an additional five days of work. Team B's MVP had three major usability issues, which took an additional two days of work. Which team was more efficient?

Team B was more efficient because it generated better-quality output and used fewer resources (time and people), resulting in lower costs.

(Assuming all engineers were involved in fixing issues, Team A worked an additional 400 hours while Team B worked an additional 128 hours to fix issues. Thus, Team A worked for a total of 2,800 hours and Team B worked for a total 2,688 hours.)

In the first month after its release, Team A's app was downloaded 10,000 times with 6,000 repeat orders from the same customers. Team B's app was also downloaded 10,000 times but with 2,000 repeat orders. (Let's assume that while most features were similar, Team A's product manager had asked the engineers to include a special filter for vegan options in its MVP.) Which team was more effective?

Team A is more effective because its users return to it more often because it has desirable filtering options. Thus, it is more likely to meet the business goals.

Note how some of these metrics are subtly tied to each other. For example, efficiently generating output inherently means you have to be productive and generate output. Similarly, efficient use of time and lower defect density can also lead to a shorter time to market and higher customer satisfaction. Thus, it is possible to optimize efficiency metrics for effectiveness.

INFLUENCING FACTORS

In Chapter 1, we discussed how you can build an effective team comprising the right number of engineers from diverse backgrounds and foster a work environment that's conducive to collaboration, innovation, and success. Team size, diversity, and various other factors can also influence the efficiency, effectiveness, and productivity measurements of software engineering teams:

Team size

The size of a software engineering team can affect its capacity to work. Large teams can complete tasks more quickly but may need additional efforts to manage communication, coordination, and decision making. Smaller teams may be slower, but they can often work more collaboratively and efficiently. Agile methodologies, like Scrum, help teams manage their workload optimally while also addressing communication bottlenecks.

Diversity

Having diverse skills on a software engineering team can improve its effectiveness and productivity. Team members with different backgrounds and expertise can approach problems from multiple angles and develop more creative solutions. However, too much diversity can also lead to communication difficulties and misunderstandings, especially when team members bring different values (*https://oreil.ly/c3aNI*) rather than different ideas to a discussion. It's necessary to create a strong sense of team and organizational inclusion to obtain the full benefit of a diverse workforce.

Role clarity

Every team member should clearly understand their role and responsibilities within the team. When everyone knows what they need to do and how they fit into the bigger picture, it can improve efficiency and productivity by reducing confusion and duplication of effort. Ambiguity about the ownership of tasks can lead to friction between team members, resulting in wasted time and effort, while clarity can lead to improved focus among team members.

Communication

Good communication implies people ask more questions and discuss more, which can improve efficiency by reducing errors and misunderstandings, and it can also enhance effectiveness by fostering collaboration and the sharing of ideas.

Work environment

The work environment can significantly impact team efficiency, effectiveness, and productivity. A comfortable, well-equipped workspace can promote focus and creativity, while a noisy, cluttered environment can be distracting and stressful. Many software companies have switched to an open office environment (*https://oreil.ly/e3H_D*) to promote collaboration between engineers. But the constant noise from phone calls, impromptu meetings, and general office chatter may not always result in productive collaboration. Therefore, it is also important to create designated quiet zones where engineers can focus in isolation for some time without any distractions.

Tools and technology

The tools and technology used by a software engineering team can affect its efficiency and productivity. Modern, reliable tools and technology can streamline workflows and automate tedious tasks, freeing team members to focus on more important work. In contrast, buggy and unreliable tools cause constant frustration, eating away at valuable time and hindering momentum. For example, imagine scrambling to meet a deadline while your testing framework throws false positive errors, requiring time-consuming manual validation.

Code health

Code health measures the quality of code in terms of maintainability, readability, stability, and simplicity and contributes to the effectiveness and efficiency of software engineering teams. *Technical debt*, or the accumulation of code problems over time, can slow development and lead to bugs and other issues requiring additional time and effort to fix. To improve code health, teams must ensure that they use clean code practices and conduct regular code reviews to identify potential issues before they become problems.

These factors can steer the software development process in different directions. Maintaining the right balance is imperative for achieving higher efficiency and effectiveness in software engineering.

Output Versus Outcome

As discussed in the "Measurement" section, efficiency and productivity measure outputs while effectiveness focuses on outcomes. It's important to note that positive outputs need not lead to positive outcomes—a highly productive or efficient team may only sometimes be effective. Understanding the difference between output and outcome will help you and your teams become more efficient, effective, and productive.

First, consider their definitions:

An *output* is a deliverable resulting from engineering tasks. An output proves that some action was taken or the team did some work.

An *outcome*, on the other hand, is the actual result of the work done. An outcome proves the team effort brought about a positive and valuable change.

To give you a better understanding of the differences between output and outcome, see Table 2-2 for some examples of outputs and their outcomes. Table 2-3 compares the impacts of outputs and outcomes.

Table 2-2. Examples of outputs and their outcomes

Outputs	Outcomes
New app released	Helped to reach more users
Code refactored	Improved code performance
A new feature added	Enhanced the user experience
Design document published	Simplified the development process
New API released	Allowed for interactions with other businesses

Table 2-3. Impacts of outputs versus outcomes

Outputs	Outcomes
Throughput Velocity Quality Capacity Code health	Business value Investment User adoption

Outputs are tangible products or services that result from a given activity or project. While necessary, they do not always reflect the ultimate goal or purpose of the project. Outcomes, on the other hand, are the changes or benefits that result from the project. These are the ultimate goals that the activity was designed to achieve.

MEASURING OUTPUTS AND OUTCOMES

Next, let's look at how outputs and outcomes are measured. If you think the metrics for output and outcome look similar to those used for measuring efficiency, effectiveness, and productivity, you would not be wrong. The metrics used for output and outcome form the core of how efficiency, effectiveness, and productivity are measured.

Output depends on activities carried out. Outputs are nearly always quantitative, with data available to show whether these have been delivered. Outputs are easy to report on and to validate. The following metrics can be used to measure outputs:

Throughput
: This is the number of items you ship to production.

Velocity
: This is the speed at which items go through the pipeline.

Quality
: This is the number of defects identified with reference to customer expectations.

Capacity
: This is the number of engineers available to work on the project.

Code health
: This is the reduction in technical debt as measured by code reviews.

Outcomes depend on the impact of activities carried out. Outcomes are more challenging to verify because they are both qualitative and quantitative. The measurement of outcomes will depend, to a great extent, on the perception of the people who receive the service. Perceptions take work to measure or report on as they are subjective and may vary from user to user based on their specific needs and expectations. The following are some of the metrics that help to measure outcomes; you will notice that they are very similar to the metrics used to measure effectiveness:

Business value
: The outcome of engineering a software product could be added revenue or cost savings for the business. Thus, this could be translated as the business value provided by the product or the value-add to the overall goals and objectives of the business that the product creates (e.g., the revenue generated by a new feature or the cost savings from a process improvement).

Investment
: This is the money that was put in to achieve the targeted outcome; for example, the budget allocated for a new product launch. A high initial monetary investment to develop a product could reduce the value of the achieved outcome.

User adoption rate
: This is the percentage of new users to all users after the activity was carried out. A high user adoption rate is a better outcome than a low adoption and contributes to the business value of the product.

Your desired outcomes will determine the outputs that are important to you. By knowing your target outputs, you can take action to meet those goals and thus your desired outcome. Take the case study of Siemens Health Services (SHS) (*https://oreil.ly/Zfcky*), for example.

SHS recognized that while it had initially embraced agile practices to improve performance, the output metrics it relied on were falling short in giving it an accurate picture of the project. It was not able to complete the planned user stories effectively by the estimated completion date. This was despite the fact that its measured velocities were reasonable.

SHS was tracking metrics that proved insufficient when trying to gauge completion dates. For example, many user stories were blocked or incomplete even when velocity rates at sprint reviews looked good. Many of the features planned for a sprint would stay "in progress" at the end of that sprint.

In response, SHS decided to shift its approach. SHS teams began by identifying their desired outcomes, or the ultimate goals they aimed to achieve. This shift in perspective led them to understand that the key to progress lay in pinpointing the outputs that truly mattered to them. They decided to focus on actionable flow metrics, such as work in progress, cycle time, and throughput.

By making this pivot toward outcome-driven metrics, SHS experienced a transformation. Its teams' efforts resulted in a remarkable 42% reduction in cycle time, signaling significant improvements in operational efficiency, quality, and collaboration. Thus, focusing on the outcomes helped SHS teams identify and measure the right outputs.

FOCUSING ON OUTCOMES OVER OUTPUTS

By focusing on achieving your desired outcomes, you can guide your project toward long-term success. Focusing on outcomes over outputs helps organizations and individuals to measure their success based on the results they achieve rather than simply the amount of work they put in. Positive results are essential for the long-term success of a project.

You can measure productivity, efficiency, and output with a variety of tools that can be integrated into your software development lifecycle. Commonly used development tools like GitHub, Jira, Trello, and Azure DevOps have built-in features that also help track and report code volume, time, progress, and defect count.

Take care not to get sucked into a productivity-measuring trap, though, where you only measure output because measuring outputs is straightforward. This approach is flawed for several reasons:

Accurate measurement is difficult

Measuring what you've produced as a team is often impossible because many people work on different tasks at different times. For example, one person might be refactoring some old code. At the same time, others work on implementing new features or fixing bugs in existing features. In this case, it's hard to know if any task was more important than another because they were all essential to building something bigger than themselves—the software product!

Excess production without value

Focusing too much on outputs can lead teams down a dangerous path where they feel obligated or pressured to produce as much as possible regardless of whether those outputs are helpful or valuable (i.e., whether they have good outcomes). Windows Vista (*https://oreil.ly/Fxnyb*) is a classic example where the focus on implementing new features resulted in software bloat and incompatibility issues with existing hardware.

Unreasonable deadlines

Setting reasonable deadlines is crucial in software projects, as unrealistic deadlines can lead to rushed and incomplete work, increased risk of errors and bugs, and low-quality results. The project team is best positioned to evaluate the requirements and should be involved in software estimation to ensure that deadlines are achievable.

One of the reasons cited for the failed launch of *HealthCare.gov* (*https://oreil.ly/lj5py*) in the US in 2013 was pressure to meet the planned schedule. Since the launch date was mandated in the Affordable Care Act, Department of Health and Human Services (HHS) employees were pressured to launch on time regardless of completion or the amount (and results) of testing and troubleshooting performed.

Burnout

The race to be more productive or efficient can lead to burnout from exhaustion. If you constantly compete with yourself or others to deliver more daily, you might overlook situations where the output is not bringing you any closer to the desired goals and outcomes. This might result in mental or physical exhaustion and even boredom with a mundane long-running project.

Focusing only on outputs can also lead to the *watermelon effect*. This term describes a situation where your outputs meet the defined targets and your metrics are "green" on the surface, but underneath, it's "red" because the outcomes or desired results have not been achieved.

Watermelon metrics (*https://oreil.ly/pJsSC*) are problematic because they give a false impression of progress or success, leading to a lack of action to improve the underlying problems. A few examples of such issues with such metrics are as follows:

- The number of lines of code written may be high, but this could also indicate inefficient coding practices that can lead to bugs, technical debt, and performance issues.
- Measuring the number of bugs fixed without considering their severity or frequency of recurrence may not provide an accurate indication of the quality of the software.
- The number of features delivered may appear positive on paper, but if they are low-value features that do not meet the customer's needs, then it's a false positive.

Measuring outcomes is more complex than measuring outputs. Business insights and user research data are needed to calculate or estimate revenue or business value. When leading an engineering team, this often means staying in touch with this research through the product team or business stakeholders involved in the project. Any disconnect in this regard can create issues for the project. To illustrate this, let me share a story.

A bright technical lead on our team, whom we'll call Brian, was passionate about the technical aspects of his role. Brian diligently ensured high-quality pull requests (PRs), clear design documentation, and robust security, privacy, risk, and mitigation plans. He excelled at the outputs of his job and could readily discuss PR counts, uptime levels, and other metrics.

However, Brian consistently struggled with outcomes. From the business perspective, the project needed to contribute to specific user priorities (outcomes). Unfortunately, it also operated in a new, somewhat ambiguous space that we were all navigating together. When our product team adjusted strategy based on research to better align with business or market needs, Brian became frustrated. He questioned whether these adjustments were truly optimal for the project's outcomes.

Brian believed the team's experience building the outputs gave team members a better understanding of what users needed as an outcome. He didn't fully grasp the business's optimization goals or the long-term implications of decisions. For example, he didn't realize that shipping the current output would effectively lock us into a two- to three-year commitment.

Despite our attempts to explain this to Brian, it became clear that he was most comfortable focusing on the outputs he could build and lead. He preferred for others, such as the product team or leadership, to consider the business outcomes and convince him of their validity.

Ultimately, Brian decided to leave the team because his beliefs were not aligned with the company's. From this experience, I now strive to clarify early in projects how they connect to the business's desired outcomes and consistently communicate this message to the engineers involved.

As a team leader, you must fully grasp the outcomes the business is interested in and communicate them clearly to your team. By focusing on outcomes, you can ensure that your team is progressing toward its goals and that resources are being used effectively. You will also be in a better position to understand what is and is not working in team structure or processes and adjust accordingly. This can lead to improved performance, increased efficiency, significant impact, and, ultimately, a more meaningful measure of success for both organizations and individuals.

In short, focusing on outcomes allows organizations and individuals to measure success more meaningfully.

Effective Efficiency

Doing things right and doing the right thing are both important. The ideal should be to do the right things right. This implies becoming effectively efficient. So, to recap, effectively efficient teams need to:

- Do things right.
- Do the right things.
- Ideally, do the right things right.

In this section, let's look at some of the steps you can take that will help you do the right things right and become effectively efficient. Naturally, since effectiveness and efficiency are different, there are some trade-offs involved. This section highlights what these trade-offs are and helps you identify and balance them.

EFFECTIVE EFFICIENCY FOR BEGINNERS

The principle of doing the right things right can apply to everyone from individuals to organizations.

An individual contributor may easily be caught up in a cycle of efficiency where they focus primarily on getting themselves up to a high level of productivity. They may invest their time in tools, techniques, and patterns that help them accomplish more to become more efficient. There may also be a tendency to take this too far and optimize code for terseness and not readability. While efficient, this may not be effective in the long run. In essence, the individual contributor would be chasing a number that indicates high efficiency but may or may not translate to better outcomes for the customers or the business.

Another common pattern I have observed among individual contributors is over-indexing on the technical side of work. In my years leading various teams at Chrome, I've seen many engineers who are technically brilliant but sometimes miss the bigger picture.

For example, Jane was a star in our Developer Experience team, known for her prowess in crafting great design docs and well-written PRs. However, she often focused narrowly on the technical aspects, losing sight of our overall business objectives.

A striking example was when she developed an advanced feature for our product portfolio that, while technically innovative, wasn't aligned with our immediate user needs. We had tried to highlight that this work wasn't a part of our plans for the year and wasn't aligned with strategy, but Jane still worked away on it.

Recognizing this gap, I introduced Jane to the broader context of our projects, pairing her with a product manager from our services infrastructure team. This collaboration was transformative. Jane began understanding the practical applications of her work, seeing beyond the code to its impact on end users and our strategic goals.

We also started a practice of having her present her technical solutions to nontechnical stakeholders across different time zones, enriching her perspective and ensuring her work was in sync with our global objectives. Jane's evolution was a testament to the importance of blending technical skills with a deep understanding of business and user needs.

It is essential that engineers understand the context of the tasks they are working on. They are then able to align their code with that context. They can judge if it is OK to compromise on code quality/abstractions in favor of readability and maintainability. Similarly, test engineers can judge which features are high priority based on the context.

Becoming effectively efficient entails more pragmatism and being aware of your context. A few ways in which developers can be effectively efficient include the following and are also shown in Figure 2-1:

Asking questions

Individual contributors must understand the larger picture before they start coding. They may wish to invest some time in unblocking themselves via research. If information is not readily available, they should ask specific questions that would help them proceed to fill gaps in their knowledge of the business domain and technical architecture.

One of my colleagues, a senior engineering manager, recommends a 20-minute rule to his team: "If you are blocked, use the first 20 minutes to do research to unblock yourself; after that, either ask our team chat or a person and let them know of the results/findings of the prior research." This enables them to better understand a problem or task and the requirements and constraints involved before asking questions, making informed decisions, and producing better-quality code. By asking questions before it is too late, developers are more likely to identify issues and potential roadblocks earlier in the development process. Similarly, testers gain a better understanding of what they should test for. This helps prevent delays and meet customer expectations at the same time.

Following standards

Developers may resort to subpar coding practices if their only aim is to ensure that the code runs (on their machines). They may hardcode parameters, not address edge cases, or skip including comments. This might produce immediate results but slow the overall development process as more time is spent on code reviews and fixing issues. Instead, making good coding practices part of the initial code delivery will lead to a slightly delayed but successful PR merge.

Collaborating

Developers and testers can communicate and collaborate with other team members even outside formal team meetings. Such exchange of ideas could provide a better understanding of the goal and create avenues to reuse tools, code, and patterns, thus driving efficiency as well as effectiveness.

Using the right tools

Developers should leverage the right tools tailored to their specific needs. For example, many developers today use Visual Studio Code (VS Code) to write and edit code. However, VS Code is much more effective when used with the relevant extensions and plug-ins that are suitable for the technology or languages being used. These extensions help developers to fine-tune workflows and optimize their productivity while producing high-quality, error-free code.

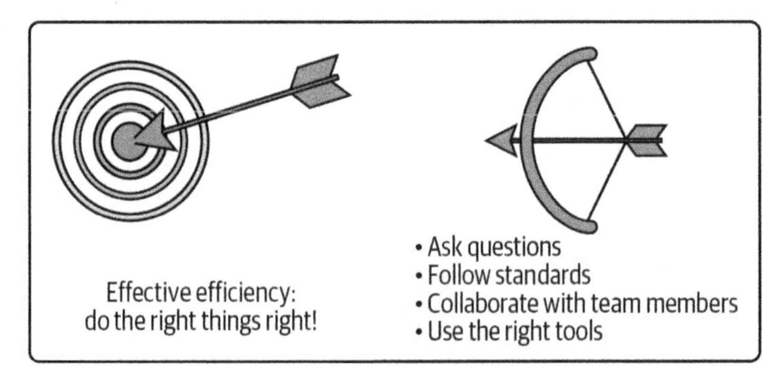

Figure 2-1. Effective efficiency: do the right things right!

These actions may seem insignificant by themselves, but asking timely questions, adhering to coding standards, promoting collaboration, and utilizing the right tools can go a long way toward helping developers become effectively efficient. These practices set the right pace and attitude necessary to streamline workflows, improve code quality, and ensure successful project outcomes.

Thus, best practices such as collaboration, following standards, and asking questions are popular because they foster efficiency and effectiveness.

MANAGING TRADE-OFFS

Although actions that promote effectiveness and efficiency may sound so wonderful that they seem to go together like peanut butter and jelly, that's not always the case. It is still possible to have too much of a good thing. There are times when something that promotes effectiveness can be detrimental to efficiency. Consider the following situations:

Choice of technology
: You want to try a new and efficient programming language or architecture pattern to accomplish a goal, but learning it will take time, and there may be unknown roadblocks. Instead, another tried-and-tested technology that the team is familiar with could help to achieve your goals more effectively.

Testing
: You might have to choose between a comprehensive testing approach that is highly effective but takes more time and resources to execute and a more efficient testing approach that may be less comprehensive but can be executed more quickly.

As a software engineering leader, you must weigh the trade-offs between effectiveness and efficiency in such situations. Several factors may influence the decisions you make to manage these trade-offs, such as:

Project timelines
: The time available to complete the project or sprints is estimated in some way before development starts, and timelines are not very flexible. Decisions about things that can affect the timeline (for example, comprehensive testing) should be factored into the estimation. If the deadline is tight despite this, then you should try to negotiate the scope of features to be developed in the sprint, prioritizing those that are most useful for users. Another option would be to plan the development by reusing components or putting in extra hours whenever possible to achieve maximum efficiency.

Budgets
: If the project has a limited budget, you may not have the time and resources to let the team members experiment. They will have to work efficiently within the budget to meet the required goals effectively.

Long-term maintainability
: Developers in your team must ensure that their code is maintainable. They may consider the trade-off between productivity and the potential technical debt that could accumulate. You should stress the importance of making the code high quality and easy to maintain in the future. This might take more time and effort, but it will be effective in the long run.

User needs
: Engineers must consider the features or components of the project that are critical to the user and focus on effectiveness, even if it means taking more time or resources.

Managing trade-offs may be challenging. With each project, you gain the experience to tackle trade-offs better to ensure that you produce high-quality code that meets the user's needs, is maintainable in the long run, and is delivered within the project's constraints. The story of Airbnb (*https://oreil.ly/qy59S*) in its early days is an excellent example of how these trade-offs can be managed effectively.

One and a half years after its launch, the founders of Airbed and Breakfast were in debt. They had around 50 visitors a day and 10 to 20 bookings, which was not enough to sustain their business. Founder Brian Chesky (*https://mastersofscale.com/brian-chesky*) had to find innovative ways to address user concerns on a shoestring budget.

The team members decided to prioritize core features that were really important to their users. From user feedback, they knew that functionalities like guest profiles with photos, listings, and booking systems provided more value to Airbnb hosts, so they prioritized these. Features like advanced search filters and personalized recommendations that personalized the user experience were tempting, but they were luxuries Airbnb couldn't afford at the time. These were put on hold.

This stripped-down approach might have lacked bells and whistles, but it possessed unyielding strength. It launched quickly, allowing Airbnb to capture early adopters and gather crucial user feedback. Slowly, feature by feature, Airbnb built upon its solid foundation, and the rest is history!

There will always be trade-offs, and learning how to manage them is an essential requirement for becoming effectively efficient.

REDEFINING TEAM PRODUCTIVITY

When striving for effective efficiency, it is important to remember that productivity is a subset of efficiency. Using modern measures of productivity that are relevant to your team can help to reduce the gap between productivity and efficiency. You can use these measures to define productivity in the context of your team.

Traditional productivity measures were designed to measure the output of repetitive tasks, such as the number of records created by a data entry clerk or the number of phone calls answered by a call center executive. These measures are easily quantifiable and may consider only a limited number of variables to account for the complexity of the work performed.

These traditional methods are not suitable for measuring productivity in software engineering, which is a knowledge-intensive activity involving problem-solving, designing, and developing complex software systems. The output of software engineering is frequently intangible, and estimating the actual value of the work done takes time and effort. For example, the value of a software system itself can be measured by the software's functionality, reliability, maintainability, scalability, and usability.

Furthermore, the work done by software engineers often involves activities that are not easily quantifiable but can significantly impact the team's overall productivity. For example, the productivity of a software engineer may be affected by the time they spend in meetings, discussions with colleagues, and learning new technologies and tools. These activities may contribute to improving the overall outcome of the project and yet may not be captured by traditional productivity measures such as throughput or velocity.

For GitHub (*https://github.com*), a leading collaborative platform for software engineering, ensuring a stellar developer experience is paramount. Developer productivity directly affects developer experience—productive developers are happy developers. To improve something, you have to measure it first. Hence, accurate developer productivity tracking, including collaboration metrics, is crucial for the GitHub team. This becomes even more vital with AI poised to revolutionize developer productivity.

Dogfooding exercises revealed that GitHub developers wished for deeper collaboration measurement. They envisioned a holistic approach that utilizes data from messaging and collaboration tools such as Slack, Jira, PRs, documents, and more, capturing both synchronous and asynchronous communication.

Productivity and collaboration metrics at GitHub (*https://oreil.ly/uDf8f*) are thus based on questions like the following:

Time to user feedback
: How quickly do developers receive valuable insights?

Asynchronous communication
: How effectively do they collaborate through non-real-time channels?

Focus blocks
: How much uninterrupted time can they dedicate to deep work?

Novel problem-solving
: How much time is invested in tackling unique challenges?

Code and bug fixing
: What's the balance between building and fixing?

Security and vulnerability
: How often are security issues encountered and addressed?

Upskilling
: During releases, how much time is spent on learning and development (self and others)?

Automated testing
: How much time is dedicated to writing and maintaining automated tests?

Context switching
: How often do developers switch tasks, and how does it impact productivity?

Subjective well-being
: How productive do developers feel?

Meeting culture
: Do developers feel that they are attending the right meetings?

These smaller metrics feed into the larger picture, helping GitHub optimize productivity, developer experience, and ultimately platform success.

This is why software engineering leaders should clearly define what productivity means for their projects without using outdated traditional measures. Instead, they should use new approaches to measure productivity in a way that considers the unique characteristics of software engineering work and focuses on measuring outcomes, quality, and value. The new productivity metrics can determine whether the team is achieving the right results for its customers and whether the team members are happy and healthy while doing it.

For optimal results, productivity metrics may be defined at the individual level as well as the project or team level.

At the individual level, teams can use SMART goals. A *SMART goal* is a goal that is *specific, measurable, achievable, relevant, and time-bound.* The purpose of the SMART methodology is to provide a template to help you write actionable, achievable goals in an organized fashion.

At the team level, there are two techniques that can help define relevant productivity metrics:

Goal-question-metric (GQM)
: The GQM approach is a methodology for driving goal-oriented measures throughout a software organization. The approach involves defining a set of goals that succinctly reflect the project's objectives and then identifying a set of metrics that can be used to measure progress toward those goals. The metrics should be specific, measurable, and relevant to the goals. This could also include relevant aspects of productivity, such as code quality, process efficiency, and team performance. You can form a holistic picture of the project by mapping business outcomes and goals to data-driven metrics. For example, consider the following metric and questions for the given goal:

Goal

Improve the search engine ranking for a website.

Questions

- What is the current ranking for the website?
- How can the ranking be improved?

Let's say you find that the ranking depends on mobile performance and user experience. You can then track the following metrics:

Metrics

- Page loading speed on mobile
- User experience metrics

Objectives and key results (OKRs)

OKRs are a practical goal-setting framework that many companies employ to align the team and individual goals with the organization's overall strategic objectives.[1] The OKR framework involves setting specific and measurable objectives and then identifying a set of key results that benchmark and monitor how to get to the objective. The key results should be specific, measurable, and time-bound. OKRs can be used to measure productivity at different levels of the organization, from individual contributors to teams and departments. Google often uses OKRs (*https://oreil.ly/kSxmL*) to try to set ambitious goals and track progress.

The GQM approach and the OKR framework can help you measure productivity in a way that is more aligned with the unique characteristics of knowledge work like software engineering. These approaches focus on measuring outcomes, quality, and value rather than just inputs and outputs. They also provide a way to align individual and team goals with the organization's overall strategic objectives, which can help improve motivation, engagement, and performance.

1 For example, the Sears Holding Company (*https://oreil.ly/Ww1eO*) released performance metrics after implementing OKRs for 20,000 associates. They found that teams consistently using OKRs were 11.5% more likely to move to the "higher performance" category. There was an 8.5% surge in hourly sales.

Do-Your-Best Goals

GQM, OKR, and SMART are based on the principle of goal setting. Research (*https://oreil.ly/jO7fu*) published in *American Psychologist* has shown that implementing specific, difficult goals consistently leads to higher performance than just urging people to do their best. Do-your-best goals have no external referent and thus are defined idiosyncratically. This allows for a wide range of acceptable performance levels. Goal specificity reduces the variation in performance by reducing the ambiguity about what is to be attained.

By redefining their productivity metrics, your team can set outcome-based goals instead of output-based goals. This ensures that your team is working on the right things and creating value for the organization, thereby becoming effectively efficient.

BALANCING EFFECTIVENESS AND EFFICIENCY

Finding the right balance between delivering outputs and prioritizing outcomes can be challenging, but the two goals need not be exclusive. Finding this sweet spot requires the definition of three crucial things:

Strategy
: Effectiveness is achieved by making improvements to the right things to achieve organizational goals. The evident first step is to identify the right things, derive the corresponding objectives, and define a strategy to achieve those objectives.

Metrics
: Once you have defined your goals and created a strategy, you need to determine how you would measure the success of your plan. As a part of this, you can determine areas where efficiency drives effectiveness.

Commitment
: Once you've decided what you want to do and how you will know you're on track, the next challenge is following through. You should plan your communication such that everyone understands organizational goals and their role in achieving them.

By using a data-driven approach to track performance metrics and committing to a strategy that balances effectiveness and efficiency, software teams can successfully balance the two factors. This can enable them to deliver high-quality code and services that meet their users' needs while operating efficiently. Netflix implemented an excellent example of this approach.

Netflix (*https://oreil.ly/N9dR9*) has always been committed to improving the video-viewing experience for its users. Its goal is for every member to view high-quality video on every device, every time. It also wants to ensure that its system can scale efficiently as the number of users grows. Netflix leveraged the extensive streaming data available to it to achieve this goal. Its strategy is to use this data to build models, algorithms, analytics, and experiments that would help it optimize video delivery. To measure the success of these models, Netflix uses metrics such as video load time, visual quality and bitrate of the delivered video, and rate of video interruptions. These are directly related to the quality of experience perceived by the user and indicate the effectiveness of the service. Thus, by continuously improving these metrics for a growing user base, Netflix has demonstrated its commitment to its goals.

EVERYDAY TIPS FOR BECOMING EFFECTIVELY EFFICIENT

Goal-setting frameworks and the use of metrics can go a long way toward making a team effectively efficient. Here are a few everyday tips that highlight everything that I have discussed so far to help leaders and teams progress in the right direction:

Prioritize effectiveness
: While efficiency and productivity are essential, they should never come at the expense of effectiveness. Develop the habit of always prioritizing doing the right thing over doing things quickly or efficiently.

Focus on outcomes
: Rather than measuring productivity based on outputs or throughput, focus on outcomes and business value. Make sure that everyone on the team understands the goals and objectives they are working toward and how their work will impact those goals.

Use metrics
: Measuring the right things can help you balance efficiency, effectiveness, and productivity. Use quantitative methods that capture your work's outcomes and business value rather than just the amount of work completed.

Foster collaboration

Collaboration can improve efficiency and effectiveness by allowing team members to share knowledge, skills, and ideas. Encourage teamwork by creating opportunities for team members to work together and share feedback. Utilize collaboration tools that also promote efficiency and effectiveness in different scenarios. For example:

- Asana, Trello, or Jira for project management
- Slack, Microsoft Teams, or Zoom for communication
- GitHub, Bitbucket, or GitLab to collaborate on code and manage code reviews

Empower team members

Empowering team members to make decisions and take ownership of their work can improve efficiency and productivity by reducing bureaucracy and streamlining workflows. When team members feel empowered, they are more likely to take the initiative and work efficiently. I shall discuss this in further detail in the next chapter.

Continuously improve

Always look for opportunities to improve efficiency, effectiveness, and productivity. Regularly review your workflows, tools, and processes, and adjust as needed to optimize your team's performance.

Empowering team members is also a great way to promote psychological safety in the team. Similarly, by using relevant metrics, you enable structure and clarity toward your goals. Thus, these everyday tips gradually take you closer to your goal of building an effective team.

Conclusion

In the preceding chapter, we explored the essential components for building an effective software engineering team: the right size, diversity, and a mindset that fosters unity and trust through communication. This chapter delved deeper into the significance of these elements and their impact on the measurement of effectiveness, efficiency, and productivity, which are the crucial metrics for gauging a software engineering team's success. By strategically balancing these factors, you can help your team drive value for both your organization and your customers.

As team leaders, you must not only understand the subtle difference between these terms but also carefully balance productivity, efficiency, and effectiveness to ensure that your teams work on the right things, correctly use the right resources, and deliver beneficial outcomes.

To achieve this balance, you must have a clear strategy, well-defined metrics tailored to your desired objectives, and a commitment to follow through. Prioritizing outcomes over mere outputs allows you and your team to truly enhance the quality of your product. Encouraging proactive risk mitigation through a focus on code health, capacity, and collaborative opportunities further solidifies this approach.

This concludes the discussion of effectiveness versus efficiency versus productivity. In the next chapter, I will introduce my enable, empower, and expand (3 E's) model for effective software engineering, which can be used to scale effectiveness for growing teams and organizations.

3

The 3 E's Model of Effective Engineering

In the previous chapters, I laid out the foundations of team effectiveness and discussed the different metrics that can be used to measure efficiency, productivity, and effectiveness. As software organizations strive to keep pace with the changing business world, teams change and evolve to become more efficient and effective. Organizations, teams, and people within those teams are going through a continuous cycle of growth and metamorphosis. We need a model for effective engineering that can stand true and help teams and team leaders through the different stages of growth.

In this chapter, I will introduce the new 3 E's model of effective engineering—a scalable model that can help engineering leaders to instill effectiveness in their teams from the ground up. (See Figure 3-1.)

In the 3 E's model, effective engineering is built through the following stages, in order of progression:

1. *Enable*

   Enable effectiveness by first defining what it means to your team or organization. The definition should explain how best to measure effectiveness in a way that makes sense for the business domain. Once effectiveness is defined, you can actively work toward it by sharing aspirations and strategies with others on the team.

2. *Empower*

   After identifying strategies to enable effectiveness, leaders must empower their teams to adopt them. Empowerment is about facilitation—removing distractions, blockers, and other complications as a leader so that teams can advance and achieve business goals effectively.

3. Expand

Expansion is about scaling effectiveness to the larger picture. It involves making teams self-sufficient so that leaders can apply their success patterns to address challenges for larger teams or at the organizational level.

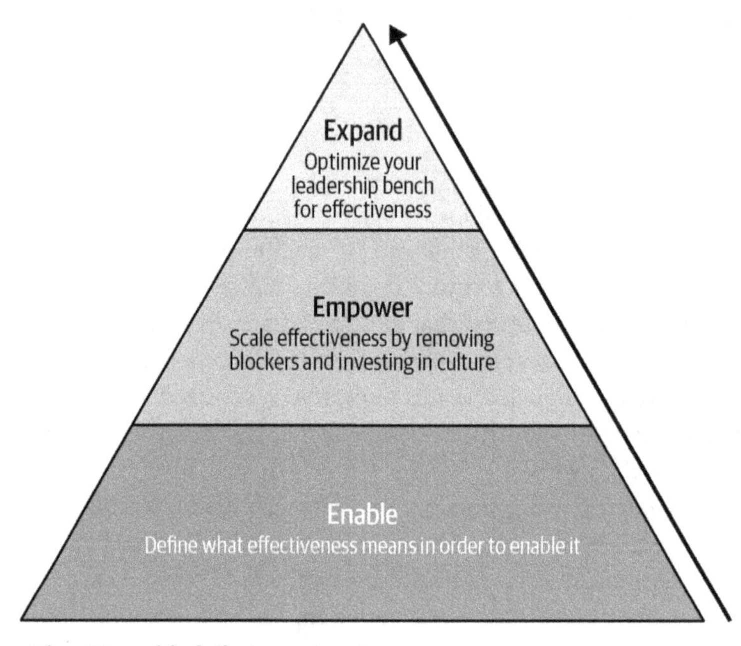

Figure 3-1. The 3 E's model of effective engineering

In this chapter, I'll focus on discussing each stage in the context of teams and team leaders. However, remember that this model can be applied on the large scale, to the whole organization, as well. If you lead more than one team, you can replicate your pattern for making teams effective once you have tried and tested it on one team. You may have to tweak your pattern for different or larger teams, but the basis of your strategy would probably remain the same based on the type of your organization. Also, remember that your team's goals should be aligned with the goals of the organization. This is why you'll see me refer to both team and organization objectives when striving for effectiveness.

Enable

To achieve success, you must know what success looks like to your organization. You must know what you're working toward. Therefore, the first thing you must do in your quest for effective engineering is define what effectiveness means for your business domain. You must *enable* effectiveness.

The Cambridge Dictionary (*https://oreil.ly/Nf_YP*) defines *enablement* as "the process of making someone able to do something, or making something possible." To make effectiveness possible within your team, you must first define what effectiveness means for your team and develop a culture of effectiveness. This requires the strategic provisioning of knowledge, support, and tools so that team members can understand and learn to practice effectiveness.

When defining effectiveness, think about what enabling effectiveness means in regards to your business domain and your type and size of organization.

DEFINE EFFECTIVENESS FOR YOUR BUSINESS TYPE AND TEAM SIZE

The previous chapter defined effectiveness as doing the right things and delivering the right outcomes. The right things can vary from one organization to another, primarily depending on the specific goals and objectives and also based on their business domain and targeted user group.

Defining effectiveness in the context of your business domain is essential because the definition can be different based on your organization's and team's goals. You shouldn't impose a blanket definition because it won't be applicable and may cause more harm than good.

Note that you could be leading a small team in a large organization or a significantly large team in a startup. In either case, a team's goals should align with or mostly derive from the organization's goals. Let's say you are leading an engineering team in a software development agency with a larger goal of customer satisfaction; your team can then have the goal of timely delivery of quality software, which aligns with the organization's goal of customer satisfaction.

Here are some general steps you can follow to define what effectiveness means for your team:

Identify your team's goals and objectives
: First identify what your team is trying to achieve. Your team's goals should be in service of the organization's goals. This could be increasing sales, improving customer satisfaction, or maximizing social impact. The SMART framework for setting goals and objectives could be used here.

Google has seen success from using the OKR framework (objectives and key results, discussed in Chapter 2) for setting goals at both the organization and team level. Not every organizational OKR needs to be reflected in every team OKR, but there should be some connection between team OKRs and at least one of the organizational OKRs.

Determine what metrics are relevant to measuring success
: Once you have identified your organization's goals and objectives, you need to determine what metrics are relevant to measuring success. For example, if your goal is to increase sales, you might track metrics such as revenue growth, customer acquisition rate, and conversion rate. Similarly, as discussed in Chapter 2, if your goal is to improve collaboration between developers and product managers, you can track the time to feedback or number of requirement gaps identified during development.

Set targets for each metric
: Once you have identified the metrics you want to track, you need to set targets for each of them. This will help you determine whether your organization is achieving its goals. For example, if your revenue growth target is 10% for the year, you can track progress toward that target. Hitting this target would be the desirable outcome for your organization.

Define effectiveness in terms of outcomes
: Finalizing targets brings you closer to understanding what effectiveness looks like for your organization or team. Your team is effective if it achieves the right outcomes that you have identified. Any practice that propels the organization toward these targets makes the organization effective.

By following these steps, you can define effectiveness for your team and create a roadmap for achieving your goals. Here are some additional tips that may be helpful when defining effectiveness for your team or organization:

Involve key stakeholders in the definition process
: This ensures that the definition is aligned with the organization's and its stakeholders' needs.

Use data and evidence to support the definition
: This makes the definition more credible and persuasive. For example, if you can show that poor performance for the app affects its usage,

then improving app performance can become a part of your effectiveness definition.

Keep the definition simple and easy to understand
: This ensures that everyone in the organization understands what it means to be effective.

Review the definition regularly and make changes as needed
: This ensures that the definition remains relevant and effective over time.

Some organizations define effectiveness in terms of productivity and efficiency. Others focus on impact and outcomes. For a software engineering organization or team, there could be many different ways to define and measure effectiveness, depending on the organization's size and type.

For most profit-driven businesses, effectiveness may be defined in terms of maximizing profitability and return on investments. Now consider that you are leading the engineering team for a crowdfunding platform that raises funds for social initiatives. The mission of such an organization would be to maximize social impact and raise awareness of issues. In this scenario, effectiveness may be measured in nonfinancial terms, such as the number of people served, the impact of the organization's services, and the level of public engagement achieved.

Even within the same organization, you could have interdepartmental targets that clash with one another. At The Telegraph (*https://www.telegraph.co.uk*), one of the UK's largest online newspapers, the advertising team wanted to increase the number of third-party ads on the website, as it would bring in more money for the organization. However, advertisements can affect a website's performance and user experience, which is bad for the core business and the engineering team's objectives. Eventually, The Telegraph found a middle ground (*https://oreil.ly/YBioq*), but it's safe to assume that such trade-offs can be expected in any similar business.

The preceding examples show that the implications of effectiveness can vary between organizations or even between teams. The conclusion is that you must define what effectiveness implies for your organization and team. To understand this better, look at Table 3-1, which shows a few common types of software organizations and what could qualify as a definition of effectiveness for them. It also shows metrics that can be used to measure effectiveness in each case.

Table 3-1. Definitions of effectiveness and metrics by team or organization type

Team/organization type	Possible definition of effectiveness	Metrics for measurement
Startup project team	Rapid delivery of minimum viable products (MVPs) to test market viability and gain user feedback	Time-to-market for MVPs
	Efficient utilization of limited resources to achieve business milestones	Burn rate and resource allocation efficiency
	Successful acquisition of funding or investment based on product potential and market traction	Amount of funding or investment secured
A team in a midsize software development agency	Timely delivery of high-quality software projects within budget and meeting client requirements	On-time project delivery rate
	Effective project management and resource allocation to optimize productivity and project success rates	Budget adherence and project profitability
	Strong client satisfaction and long-term partnerships	Client satisfaction surveys and client retention rate
Open source community project	Active community engagement and participation, including contributions, feedback, and collaboration from community members	Number of active contributors and contributions
	Wide adoption and use of the software within the target user base	User adoption and download statistics
	Continuous development and improvement based on community-driven feedback and contributions	Release frequency and community-driven feature requests implemented

INITIALIZE EFFECTIVENESS

Once you have defined what effectiveness means for your team, you must share it clearly with your team members so they know what you mean when asking for effective work.

There are several ways to share definitions of effectiveness so that team members understand and adhere to them. Some of the most useful methods include the following:

Communication

Communicate the definitions of effectiveness to team members through various channels, such as team meetings or email. Ensure the definitions are clear, concise, and easy to understand, especially for the engineers to whom the definitions apply.

Training

Provide team members with training on the definitions of effectiveness. This training should cover the meaning of these definitions and how they can be applied to their work. Teams can tap into organization-level training, if available; team leaders can ensure that their team attends these.

Measurement

Track team members' progress against the definitions of effectiveness. This will help ensure that they are aware of their performance and working toward the organization's and team's goals.

Feedback

Provide team members with feedback on their performance against the definitions of effectiveness. This feedback should be constructive and helpful and must be provided on time.

Rewards and recognition

Recognize team members who meet or exceed the definitions of effectiveness. This will help to motivate employees to continue to perform at a high level.

By using these methods, leaders can ensure that their vision and goals have reached everyone on their team and they are sufficiently motivated to work toward them. In general, members of a software engineering team must realize that being effective does not mean they care more about using a new technology or approach, but instead, it means they care about the value it adds for their customers. Being effective implies they should strive to understand the problem being solved and build solutions that delight customers.

In addition to these foundational steps to initialize effectiveness, the 2020 Gartner Software Engineering Team Effectiveness Survey (*https://oreil.ly/mWUjs*) reveals three crucial enablers that can equip software engineering teams to effectively deliver on stakeholder value and responsiveness goals. You can consider fostering these three factors in your team. The survey suggests focusing on empowerment, critical skills, and servant leadership to build substantially higher team performance.

Empower teams to shape standards

Software engineering standards can be restrictive, preventing teams from achieving broader business objectives. To optimize the benefits and limit the restrictions, Gartner recommends that leaders must allow software engineering teams to take part in creating standards that work best for them. The result makes them 23% more effective than their counterparts who don't participate in standard setting. Teams should also evolve the standards as technology and business requirements change and provide recommendations for situations where standards don't apply. They should address user experience, architecture, database design, and integration standards, as these issues strongly influence team effectiveness.

Promote critical skills

Effective software engineering leaders ensure that their teams possess the necessary skills and competencies to independently achieve their objectives, resulting in improved outcomes and reduced delays. They prioritize the skills and competencies required for successful day-to-day workflows while fostering versatility among team members, enabling them to contribute to different activities. According to Gartner (*https://oreil.ly/cC1vW*), teams comprising versatile members are 18% more effective than teams composed solely of specialists. Encouraging team members to assume new roles outside their current expertise is one approach to fostering versatility.

Practice servant leadership

When team members are burdened with administrative tasks, their ability to engage in value-adding activities is diminished. Leaders can assume these responsibilities to make the team more effective. For instance, when leaders proactively identify and address obstacles, their teams experience a 16% boost in effectiveness (*https://oreil.ly/skeX5*). Additionally, when leaders coordinate with stakeholders such as project managers or governance partners, team effectiveness increases by an additional 11%.

Thus, leaders can go the extra mile by removing obstacles for their team members, encouraging them to be versatile, and letting them own the standard for the project.

Defining and initializing effectiveness, combined with the advice shared in the previous chapters (such as building a strong team with a sense of psychological safety and team spirit), can help accomplish the first E (enable) of the 3 E's model of effectiveness. Let's now look at how you could empower effectiveness within your team or teams.

Empower

If enabling is about identifying what needs to be done to become effective and initializing it, empowerment takes you a step further. *Empowering* effectiveness involves providing the necessary support, resources, autonomy, and opportunities for growth and development. Empowering grants individuals or teams the authority, independence, and trust to take ownership of their work and make decisions. It emphasizes giving individuals the power, confidence, and freedom to act independently and be accountable for their actions and outcomes. Empowering focuses on cultivating self-belief, fostering growth, and promoting ownership.

In this section, I will share methods of empowering effectiveness that I have either developed or learned from others over the years, as shown in Figure 3-2.

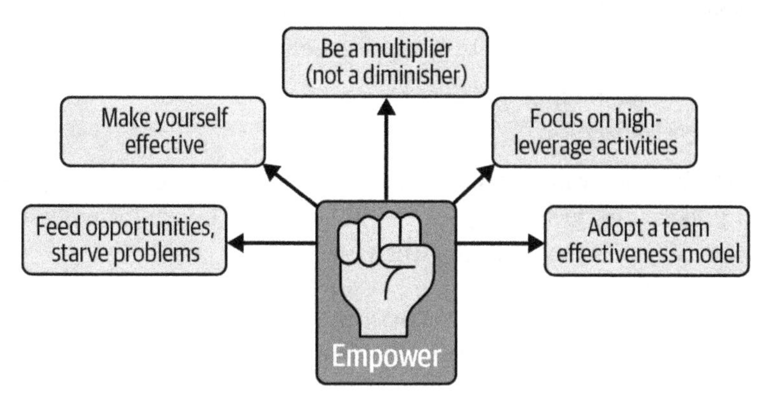

Figure 3-2. Empowering effectiveness

FEED OPPORTUNITIES, STARVE PROBLEMS

My core principle for empowering effectiveness is to feed the opportunities and starve the problems.

Feeding opportunities involves creating an environment where individuals and teams can thrive by providing them with the resources, support, and opportunities necessary for growth and success. It focuses on maximizing potential and enabling individuals to reach their full capabilities. It involves allowing individuals to engage in activities, projects, or roles that align with their strengths and interests. Examples of feeding opportunities include the following:

- Assigning a detail-oriented individual to work on quality assurance or assigning a creative thinker to handle design-related tasks
- Offering training, workshops, or mentorship programs that allow individuals to enhance and refine their strengths, whether they be technical skills, leadership development, or domain-specific expertise
- Allowing team members to leverage each other's strengths by forming cross-functional teams where individuals with complementary strengths work together on a project

Starving problems involves minimizing the impact of obstacles, challenges, or inefficiencies that can hinder effectiveness. It focuses on addressing and resolving issues swiftly to maintain productivity and positive momentum. A few scenarios where you might have to starve problems include the following:

- Identifying and addressing bottlenecks or inefficiencies in existing processes by conducting regular process audits, gathering feedback from team members, and implementing improvements based on the findings
- Promptly addressing conflicts or issues within teams to minimize their negative impact
- Ensuring that teams have the necessary resources, such as adequate staffing, tools, and technology, to accomplish their tasks efficiently; this prevents unnecessary delays and frustration

Here are a few more related techniques that can help software engineering leaders to feed opportunities or starve problems:

Continuous delivery and feedback loops
: Teams implementing continuous delivery pipelines are able to release code and features more frequently, creating quicker feedback loops. This allows them to leverage the opportunities created by real-world user data and insights. They are able to rapidly identify and address problems, thus becoming more effective.

Kanban boards

Use of Kanban boards in software projects helps to visualize the flow of work. Kanban is based on the principles of "just-in-time" (only at the right time) and "pull," which means that tasks are performed only if there is a demand for their execution. Kanban boards allow you to divide tasks into stages and visually track their completion. By limiting work-in-progress (WIP) tasks and focusing on completing existing tasks before taking on new ones, using Kanban boards effectively "starves" the potential problems of overloaded team members and unfinished work. This is a great strategy for use in an uncertain project environment.

Delegation and empowerment

Managers who delegate tasks effectively and empower their teams to make decisions are "feeding" opportunities for individual ownership and growth. This motivates team members and reduces bottlenecks caused by micromanagement. By "starving" the problem of centralized control and decision making, managers can unlock the full potential of their team and achieve better results.

Engineering teams are better off channeling their efforts toward areas that have the potential for growth, innovation, and positive impact. Let's say users of your web application have encountered performance issues on a specific web page. Your team identifies that the issue is caused by a specific database query. Team members proceed to optimize the web app's performance by indexing tables referred to in the query and making modifications to create a better execution plan, then test and implement the changes. The performance issue is resolved, and everyone is happy.

However, this is a great opportunity asking to be fed. The same optimization approach could be used for other queries on other web pages that can potentially have similar issues in the future. This would help the team enhance the overall performance of the system. By allocating resources, time, and attention to optimizing the application's performance, you can not only address the existing problem but also position yourselves for future success. This is a forward-thinking approach that aligns with the idea of feeding opportunities (performance optimization) rather than exclusively fixating on the problem (poor performance).

By feeding opportunities and starving problems, leaders can create an environment that fosters growth, development, and high levels of effectiveness. It involves nurturing individuals' potential while addressing obstacles that hinder their progress, ultimately leading to improved outcomes and success.

STRIVE FOR INDIVIDUAL EFFECTIVENESS

Effectiveness does not imply getting more things done by working longer hours. Effective engineering teams are the ones that get things done efficiently and that focus their limited time on the tasks that produce the most value. This requires that everyone on the team, including those in leadership roles, start by making themselves effective. To empower your team to be effective, you must also lead effectively. I discussed some strategies on how to cultivate effectiveness as a leader in Chapter 1. To build on that, let me start with a story to illustrate why individual effectiveness is important.

I've navigated the challenges of scaling teams quite a few times, but a pivotal moment in this journey was when my team's growth began to overwhelm my capacity to manage every detail. We had significantly expanded the number of projects and people that I was leading, introducing not just coordination headwinds but also a lot of context-switching and knowledge that was becoming increasingly hard to keep in my head.

In the early days of our expansion, my inbox was a never-ending stream of decisions waiting for my input. This deluge wasn't just exhausting; it was counterproductive. I realized I was the bottleneck stifling our progress. The first step to solving any problem is acknowledging it, and that's what I did. After that, it was about finding a way out using the following essential strategies:

Using delegation as a tool, not a retreat

Delegating wasn't simply about offloading work; it was about empowering others. I transitioned from being the sole decision maker to being a mentor and guide. I started by identifying routine decisions that could be managed by team leads. This wasn't a one-off process. It required regular check-ins and adjustments to ensure everyone was comfortable and effective in their new roles. I collaborated with my leads on putting together a delegation plan for decision making, including starting to use RACI matrices (which I'll discuss in Chapter 4) more often, which helped.

Building a culture of trust and transparency

Delegation can't work in a vacuum. It needs a supportive environment. We cultivated a culture where transparency was paramount. Regular open forums and team meetings ensured everyone was aligned and aware of ongoing projects and challenges. Trust was another cornerstone. I trusted my team members to make decisions, and they trusted me to support them, even when mistakes were made. This mutual trust was crucial for our collective growth.

Using process optimization beyond command-and-control
: The traditional command-and-control model wasn't suitable for our dynamic environment. We adopted agile methodologies, emphasizing flexibility and continuous improvement. We also embraced tools and practices that facilitated remote collaboration, which was crucial for our globally distributed team.

The most profound impact of this shift was the growth in leadership capabilities across the team. As team members gained confidence in decision making, they also developed a deeper understanding of the broader business context.

With the team more autonomous, my role evolved. I moved from firefighting daily issues to focusing on strategic objectives. This included exploring new technologies, fostering partnerships, and planning long-term product roadmaps. My role became more about enabling the team's success rather than micromanaging its tasks.

This journey wasn't without its challenges, but the transformation was remarkable. My team evolved from a setup where capabilities were concentrated at the top to a more resilient and dynamic structure. The team's agility and ability to respond to rapid market changes improved significantly. Most importantly, we cultivated a generation of leaders who would go on to drive innovation and excellence in their respective areas.

While this personal anecdote sheds some light on effective leadership in a specific context, you can also learn much from Peter Drucker's *The Effective Executive* (Harper Business Essentials, 2006) about consciously making yourself effective irrespective of what kind of work you do. This book prescribes the following habits (*https://oreil.ly/mvscQ*) you can cultivate and apply in most organizations, including software engineering teams:

Know where your time goes
: As a manager, it's crucial to understand how you're spending your time and ensure that it aligns with your goals and areas of impact. While managing emails and attending meetings are essential parts of the job, they shouldn't consume a disproportionate amount of your time. Identify the

activities that contribute most to your core responsibilities and the areas where you want to be impactful. This awareness will help you prioritize effectively, delegate tasks that can be handled by others, and allocate your time and resources to high-impact initiatives. Understand your work habits and optimize your schedule to maximize productivity during your peak hours. Regularly assess if your time is being spent on activities that drive the most value for your team and the organization.

Focus on what you uniquely can contribute to your organization
: As engineers and leaders, you must identify your unique skills, expertise, and knowledge. Determine the areas where your contribution adds the most value and allocate your time and energy accordingly. For example, if you excel in frontend development and user experience design, focus on contributing your expertise to enhance the user interface and overall user experience of your organization's software products. Maximize your impact by leveraging your strengths and providing valuable insights into your area of specialization. Be yourself. Don't try to be something you're not. Discern patterns from your past performance. Ask yourself, "What are the things that I seem to be able to do with relative ease that come rather hard to other people?"

Build on your own strengths, the strengths of your colleagues, and the team
: Recognize that effective software engineering and leadership require collaboration and building on the strengths of individuals and the team. If you are a leader, lead from your strengths. Do not concern yourself with what you cannot do. Focus all of your energy on what you can accomplish. Identify the strengths of your colleagues and team members and encourage them to develop and utilize those strengths. Create a culture that fosters knowledge sharing, collaboration, and continuous learning.

Concentrate on a few major areas where superior performance will produce outstanding results
: The secret of effectiveness is concentration or doing one major thing at a time. In the fast-paced and dynamic field of software engineering, it's essential to identify the critical areas where your expertise can have the most significant impact. Concentrate on those key areas where exceptional performance can lead to outstanding results. This may involve focusing on critical projects, technologies, or processes that align with the organization's goals and objectives. For example, if your organization is launching a

new mobile application, concentrate on delivering exceptional performance while designing and implementing key features, thus ensuring a superior user experience. By focusing on these major areas, you can contribute to the project's success and drive outstanding results. At the same time, you must ruthlessly prune the activities that should do well but, for some reason, do not produce results.

Make effective decisions

Decision making and managing trade-offs are inherent to software engineering and leadership. Ensure that you gather relevant information, analyze options, consider potential risks, and make informed decisions. Embrace a data-driven and analytical approach, seeking input from team members and stakeholders when necessary. Continuously evaluate the outcomes of your choices, learn from them, and refine your decision-making process. For instance, when choosing between different technologies for a specific project, gather relevant information, evaluate the pros and cons of each option, and consider factors such as scalability, maintainability, and compatibility. Make an informed decision based on your analysis and the project's requirements.

Through strategic time management, focusing on strengths and collaboration, and making informed decisions that align with the organization's goals, software engineers and leaders can enhance their effectiveness, drive impactful outcomes, and contribute to the success of their organizations.

While these takeaways from *The Effective Executive* provide a reliable guide to empowering effectiveness among individuals, let's look at some models that can be useful to empower a team's effectiveness.

FOLLOW TEAM EFFECTIVENESS MODELS

Team effectiveness models are frameworks or theories that provide a systematic approach to understanding and assessing the factors contributing to high-performing teams. The frameworks enable you to debug better and understand what's working (or not) on your team, what team members are experiencing, and how you can empower their best work. These models guide teams in improving their effectiveness and achieving their goals. By choosing to follow one of these models, you can empower your team with a tool to track and improve its effectiveness. Let's explore these models here.

Lencioni's model

Patrick Lencioni's model (*https://oreil.ly/LUUhC*) focuses on the five dysfunctions that can hinder team effectiveness. The Lencioni model is a framework for diagnosing and treating organizational dysfunction. It was first published in 2002 in a book called *The Five Dysfunctions of a Team* (Jossey-Bass). The Lencioni model is structured like a pyramid, with "absence of trust" at the bottom of the pyramid and "inattention to results" at the top, as shown in Figure 3-3.

Figure 3-3. The Lencioni model

The following is a list of dysfunctions that can plague a team according to this model:

Absence of trust
: When trust is absent among team members, they are unable to be vulnerable with one another or show weakness. As a result, team members will not admit mistakes and seek help from others.

Fear of conflict
: When a team lacks trust, its members fear conflict and keep quiet during debates. This results in poorer decisions than if their opinions had been voiced openly.

Lack of commitment
: Fear of conflict leads to indecision and poor cohesion among team members. As a result, there is little commitment to decisions, which creates an environment where ambiguity prevails.

Avoidance of accountability
: A lack of commitment leads to a breakdown in accountability among team members. If one has not bought into the decision, they will be less likely to hold peers accountable for their actions.

Inattention to results
: If team members don't feel accountable for their actions and goals, they tend to put the needs of individuals above those of the group. This can lead to miscommunication within teams and a decline in company performance overall.

Dysfunctions such as power struggles and lack of trust between team members are common in many organizations, including software engineering teams. For example, you often see groups of developers and testers working as opponents in a bug-finding contest instead of as a team. This can lead to both irrelevant bugs being reported by the test engineers and superficial fixes being applied by the developers.

According to Lencioni, overcoming the absence of trust requires the team to show vulnerability and a willingness to take risks. Leadership should acknowledge dysfunctions, seek to understand their root causes, and then work together with other leaders on a plan for addressing them that will result in more productive teams. In the context of the developers-versus-testers scenario mentioned earlier, this could mean conducting peer reviews for test cases and fixes and ensuring that everyone on the team focuses on the customer requirements over their preconceived notions.

The Lencioni model is a great tool for identifying what's going on in your organization. Using the model can be a helpful way to identify how you and your team can work together more effectively. If there's a gap between where you are now and where you want to be, it's important to address that issue head-on.

Tuckman's model

Developed by Bruce Tuckman in 1965, Tuckman's team development model (*https://oreil.ly/Xae9N*), shown in Figure 3-4, describes the stages teams typically go through as they develop and mature:

Forming
: This is the stage where your team is just coming together. Each team member feels things out during this stage to gauge how the team will work.

Storming
: The team is still getting to know each other. But as they become more familiar, power struggles and relationships can emerge.

Norming
: The team finds a relative groove. They better understand how the team works (and how you, as the team leader, manage the team).

Performing
: The performing phase is where the magic happens. At this stage, your team members trust and respect each other.

Adjourning
: The final stage occurs when the team disbands after completing its objectives. In this stage, the team members reflect on their achievements, share lessons learned, and celebrate their success before disbanding or moving on to new projects. It's important for leaders to facilitate a smooth transition, ensure knowledge transfer, and provide closure for the team members.

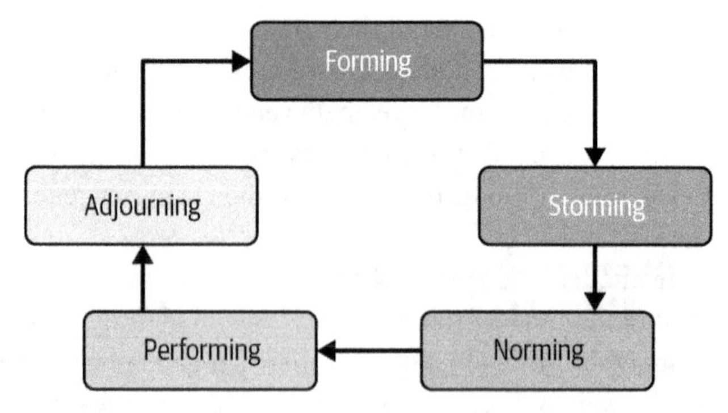

Figure 3-4. The Tuckman model

You would typically experience all of these stages in a software engineering team as the team starts to work on a new project. As team members get to know and trust each other more, they progress through each stage—and become stronger and more effective throughout the progression. However, it's important to remember that Tuckman's team development model isn't necessarily linear; you might bounce between phases. For example, you may be riding high in the performing phase only to fall back to the storming phase after introducing new members to your team. As a team leader trying to empower yourself and your team, you must know what to expect at each stage and be prepared to guide your team through it. For example, if you expect a storming phase, you can prepare yourself better by organizing team-building sessions to help the team get acquainted with each other and new team members.

Lencioni's model focuses on overcoming dysfunctions that hinder team effectiveness, while Tuckman's model describes the stages of team development. Both models provide valuable insights into understanding and improving team dynamics, empowering teams to reach higher levels of effectiveness and achieve their objectives.

Team effectiveness models are helpful tools, but they will only transform your team after a while. These models are not a panacea. You need to adopt them and apply them consistently. So be patient with yourself and your team. The idea is to *pick a model and stick with it*. In addition to these models, let us look at other ideas on team effectiveness that would be useful to any team.

MULTIPLY EFFECTIVENESS

When Taylor Murphy started managing the data team at GitLab (*https://about.gitlab.com*), he had not anticipated the fast-paced growth that he encountered at the startup. The data team and its effectiveness had to grow to keep pace with the business, which meant Murphy had to level up his game. The following are some strategies (*https://oreil.ly/mBFgc*) that helped:

Using the right tools and processes

Find the best tools for your team to empower it. For example, Murphy described a case in which the team had to upgrade its data warehouse and analytics solution as the size of data grew. A few changes that proved extremely valuable were as follows:

Adopting dbt

This data build tool streamlined data workflow and eliminated repetitive tasks, allowing the team to do more with less.

Moving to Snowflake
: Upgrading from Postgres to a more scalable data warehouse enabled efficient handling of larger datasets.

Investing in documentation
: Well-documented processes and onboarding materials empowered new team members and reduced friction.

Reducing meetings and protecting his team
: Some meetings are necessary for intrateam collaboration, but others need only one or two representatives from the team to join. Murphy tried to shield direct reports from unnecessary meetings as much as possible to maximize their productive time.

Securing executive buy-in and resources
: If you are a team leader or manager supporting a specific function, having someone on the executive team who understands the value of the function can be a big boost for your team. Resources will only be sanctioned if this kind of buy-in exists. Based on his experience, Murphy recommends advocating for at least a director-level leader who can support your function.

Additional tips
: Here are a few other pointers that helped to improve effectiveness at GitLab:

  - Hiring well and investing in employee development
  - Planning for growth and scaling the team effectively
  - Focusing on core business value rather than reinventing basic tools
  - Being mindful of your own career path and priorities

By implementing these strategies, Murphy successfully multiplied the effectiveness of his small data team, enabling it to handle the demands of a rapidly growing company. His learnings offer valuable insights for anyone managing a data team or seeking to boost team productivity in general.

Like Murphy, leaders play a crucial role in shaping team dynamics. If done right, the act of empowering the team can multiply its effectiveness.

Some leaders can increase effectiveness, and some decrease it. This idea is best captured in Liz Wiseman's book, *Multipliers: How the Best Leaders Make Everyone Smarter* (HarperBusiness, 2017).

Multipliers introduces the idea that some leaders act as "multipliers" while others unintentionally act as "diminishers." *Multipliers* are leaders who amplify the intelligence, capabilities, and contributions of those around them. They create an environment that encourages and empowers team members to excel, bringing out the best in people. These leaders effectively leverage the team's collective intelligence, facilitating learning and promoting growth. Thus, multipliers are leaders who provide the backbone for team members to realize their full potential while collaborating with each other, thereby empowering them to multiply their effectiveness.

On the other hand, *diminishers* are leaders who unknowingly stifle the capabilities and potential of their team members. They tend to hoard power, make all the decisions themselves, and undermine the confidence and autonomy of their team. Diminishers may unintentionally create a dependency culture and limit their team members' growth and performance.

It is easy to spot multipliers among software engineering leaders. Such leaders demonstrate trust by delegating some of their responsibilities. For example, a leader could involve a team member when discussing requirements with stakeholders or let them lead a meeting. Team members led by multipliers feel safe contributing ideas or asking questions about the projects they are working on. They feel encouraged to learn and grow within their teams. They feel a sense of ownership toward their work that motivates them to be effective. Gradually, the team becomes more and more effective in delivering the right outcomes.

IDENTIFY HIGH-LEVERAGE ACTIVITIES

A few years back, during a collaboration with YouTube on improving the accessibility of their product, Google teams tackled a significant hurdle: ensuring the website worked flawlessly for users with visual impairments. Screen readers are crucial for such users, and we realized that they behaved differently across desktop and mobile devices, creating a testing nightmare.

Our traditional testing methods weren't cutting it. To truly understand the issue, we embarked on deep research into screen reader behavior on various devices. This investigation led us to a breakthrough: a comprehensive testing framework.

This framework, including a custom automated testing library and heightened team awareness of screen reader nuances, revolutionized our release workflow. Subsequent releases were significantly faster and boasted far fewer accessibility issues. By tackling a single pain point, we not only solved a recurring accessibility problem but also boosted our overall productivity. This project is a shining example of what managers usually refer to as a *high-leverage activity*.

In his book *High Output Management* (Vintage, 1995), Andy Grove, former CEO of Intel, shares his perspectives on management. He describes how leaders can manage their teams more efficiently with maximum leverage.

Leverage, as Grove defined it in the book, is the output generated by a specific type of work activity. As such, an activity that has high leverage will generate a high level of output and vice versa. High leverage is the ability to do more with less effort, time, or money.

There are two types of leverage: personal and organizational.

Personal leverage refers to the tools, techniques, and knowledge you have that allow you to do more with less effort. *Organizational leverage* refers to teams and structures that enable you to accomplish more with less time or money. Organizational leverage is often overlooked, but it's one of the most important components of building a high-output management team.

If you improve your personal effectiveness, maybe you can make things 10% better. If you improve the effectiveness of an organization of 30 people by 10%, then you've effectively added three software engineers.

Software engineering team leaders can apply the high-leverage principle from *High Output Management* to enhance their effectiveness and drive team productivity. Leaders should prioritize activities that improve the performance of their team members. They could start by providing mentorship, coaching, and skill development opportunities to enhance the skills and capabilities required for a project. Delegation of tasks and responsibilities can empower team members to take ownership of those tasks. They can help the team identify the high-impact activities for specific projects and prioritize them to increase the productivity and effectiveness of the team.

LESSONS FROM GOOGLE

At Google, many aspects of the company's engineering culture have empowered Googlers to become effective. At an organizational level, it's essential to have a culture that your team values and is willing to defend. A strong focus on innovation, collaboration, and quality characterizes Google's engineering culture. Engineers at Google are given a great deal of autonomy and are encouraged to take risks and experiment. The company also invests heavily in training and development so that engineers can stay up to date on the latest technologies. Some key features of Google's engineering culture that foster effectiveness are as follows:

Standardize and share

Solving a problem well once and then getting everyone internally to adopt it has had huge payoffs at Google. We share tools, abstractions, and conventions: from the early days, Google has invested heavily in tools and abstractions like shared libraries, tools, protocol buffers, MapReduce, Bigtable, and more used throughout the engineering organization. Each team spends fewer mental cycles choosing which tools to use, dedicated tools teams can focus on improving engineering productivity, and those improvements quickly propagate to everyone already using the tools or services. When contrasted with engineering organizations where each team might use vastly disparate toolchains, this philosophy also means that it's much easier to understand the designs behind many projects once you've learned the fundamental building blocks.

Reuse

One reason teams quickly become productive within Google is that the company has invested so many resources in creating reusable training materials. These have helped my teams ramp up on Go, TypeScript, and internal frameworks. The training materials covered the core abstractions at the company; code labs highlighted relevant snippets of the codebase and the lessons folks learned. Without these training materials, teams would have taken much longer to learn about the many technologies needed to be effective, and it would have meant that teammates would have had to spend more time explaining them to me.

Automate the right things

Automation is a force multiplier, not a panacea. Automation helps Googlers to become more effective by improving efficiency, reducing costs, improving quality, and increasing safety. Automation powers many areas in Google, from quality control to data centers. Google's data centers are some of the largest and most complex in the world. They require a massive amount of energy and resources to operate. Google uses automation to help manage its data centers, which helps to keep costs down and improve efficiency.

These are just a few of the ways in which Google's culture contributes to its scaling effectiveness. The company's focus on standardization, sharing, reuse, and automation has helped create a supportive work environment for its engineers, enabling them to be effective.

In summary, to empower team effectiveness, leaders can provide their teams with the platform to become effective by creating a culture that fosters trust and collaboration. At the same time, individual team members must also become effective in their everyday work by using their time wisely and focusing on the most critical goals of their teams. When teams are successful, it can set the stage for the organization's leaders to replicate their success and expand effectiveness across the board.

Expand

Expanding effectiveness refers to increasing an organization's overall effectiveness, efficiency, and success across its various departments, teams, and functions by making them self-sufficient. The focus here is on growing leaders within the organization, who can then lift up their teams to make them more effective. On successful projects, as the team size grows, the scope of a leader's responsibility grows. A leader who manages one team may rise to lead a team of teams or a department. Such a leader must now apply their successful effectiveness patterns across the growing team or multiple teams. Failure to do so can lead to problems, as in the case of my colleague Cathy, a seasoned manager who struggled with scale.

Cathy had been with Google for several years, managing various teams within Chrome. She was well respected, knowledgeable, and had a good number of direct reports. However, as the scope of her responsibilities grew, Cathy began to struggle with scaling her management effectiveness.

The issue became apparent during a major project overhaul. Cathy was trying to stay intimately involved in every aspect of her team's work, from detailed code reviews to strategic planning meetings. While her intentions were good, this level of involvement was unsustainable and started to create bottlenecks.

As Cathy's manager, I worked with her to address the situation by using the following methods:

Empowerment through trust

The first step was encouraging Cathy to trust her team more. This meant letting her direct reports take more ownership of their projects. She needed to step back from being deeply involved in every decision and instead focus on guiding and advising.

Effective delegation

I worked with Cathy to identify which tasks she could delegate to her team leads. This process involved recognizing which decisions required her expertise and which could be effectively handled by her capable team.

Streamlining communication

With numerous reports, communication was a challenge. We implemented structured communication channels and regular, but concise, update meetings within Cathy's team. This approach ensured that Cathy stayed informed without getting bogged down in the minutiae.

Fostering a culture of autonomy

Cathy started to cultivate a culture where her team members felt empowered to make decisions within their domain. This shift not only reduced her workload but also boosted her team's confidence and sense of ownership.

Setting priorities and boundaries

We helped Cathy redefine her priorities. She began focusing more on strategic planning and less on operational details. Setting clear boundaries for her involvement in different projects helped her manage her time more effectively.

Mentoring and developing leaders

Recognizing the potential in her team, Cathy started investing time in mentoring her direct reports, preparing them to take on more leadership responsibilities. This approach helped her build a strong second line of leadership within her team.

Reflective practices

Cathy started a practice of regular reflection on her management style, seeking feedback from her team and peers. The self-awareness she developed from this was crucial to her ongoing development as a leader.

Over time, Cathy's new approach paid off. The team operated more efficiently, with less reliance on her direct involvement in every detail. This change allowed her to focus on broader strategic goals, contributing more significantly to the project's success.

To me, this incident was a crucial lesson in software engineering leadership, especially in dynamic and fast-paced environments. Experienced managers, no matter how competent, must continuously adapt their leadership style to meet the evolving needs of their team and projects.

Scaling oneself isn't just about managing more work; it's about enabling the team to work smarter and more autonomously. I will talk about my strategy to expand effectiveness later, but first, it's necessary to fully grasp the challenges leaders encounter as they grow.

LEADERSHIP CHALLENGES

The leader's role transforms as their domain expands. They are faced with new challenges (Figure 3-5). With great power, you not only get great responsibility but also more complexity. As you transform as a leader, you shift your focus from individual technical expertise to people and broader organizational considerations. The following stand true as you rise further as a leader:

It's more about people
: As your domain expands, your attention increasingly centers on people. Effective leadership involves developing and nurturing a high-performing team, creating a positive work culture, communicating your vision, and fostering collaboration. Leaders must focus on building strong relationships, understanding and addressing the needs of employees, and empowering them to achieve their full potential. Leaders become more involved in talent management, coaching, and ensuring the right people are in the right roles.

It's less about technical expertise
: In the early stages of your career, you will often rely on your technical knowledge and hands-on involvement to drive success. However, as your responsibilities grow, you will have to shift your focus away from personal technical proficiency. Instead, you must delegate technical responsibilities to subject matter experts and trust their team members' capabilities. Effective leaders recognize that their role is to provide guidance, set direction, and create an environment that enables others to excel.

Your domain becomes broader, making you even more removed
: As your domain expands, your responsibilities become more expansive and complex. You must oversee multiple functions, departments, and teams, focusing on aligning efforts with common goals. This broader scope necessitates a more strategic and visionary approach to leadership. Leaders must make decisions that consider the organization's overall health,

financial sustainability, market positioning, and long-term growth. They may become more removed from the day-to-day operational details but remain accountable for the organization's success.

There are more distractions and complications

With growth comes increased distractions and complications for leaders. You'll face challenges such as managing increased organizational complexity, adapting to market dynamics, navigating regulatory requirements, and addressing stakeholder expectations. You must develop effective time management skills, delegation abilities, and the capacity to prioritize tasks significantly impacting the organization's success. You must become skilled at managing ambiguity, making tough decisions, and balancing competing demands.

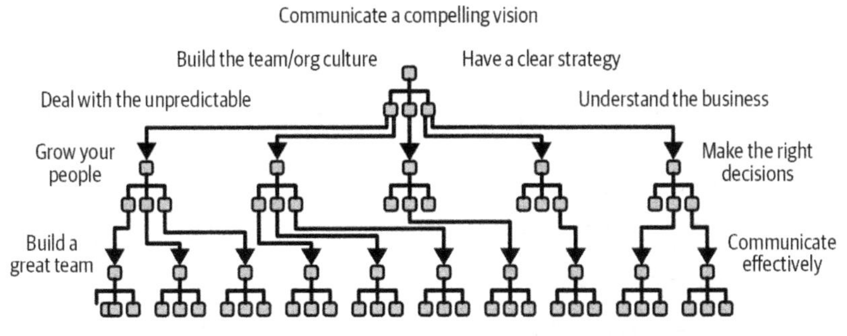

Figure 3-5. Your leadership role changes as your domain grows and distractions and complications multiply

In short, you face all the challenges you envisioned as a leader but with an increased magnitude as you grow within your organization. But as they say, leadership is an art. Let's see what it takes to master this art.

THE THREE ALWAYS OF LEADERSHIP

Leadership often includes a combination of strategic thinking, emotional intelligence, effective communication, and the ability to make tough decisions. Balancing these elements with grace is indeed an art. In Chapter 6 of *Software Engineering at Google* (*https://oreil.ly/DnNyv*) (O'Reilly, 2020), Ben Collins-Sussman talks about the natural progression from leading one team to leading a set of related teams and techniques that can help you stay effective as you continue growing as an engineering leader. These techniques highlight a principle called the three always of leadership, illustrated in Figure 3-6.

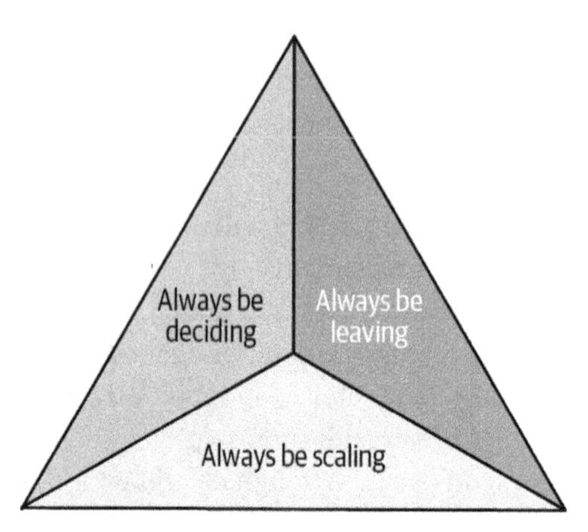

Figure 3-6. The three always of leadership

The mantra of "always be deciding, always be leaving, always be scaling" is a reminder that leadership is a constant process of change and adaptation. Leaders must be able to make quick decisions, even when they don't have all the information they need. They must also be willing to let go of tasks others can do. Finally, they must constantly scale themselves to lead a growing organization effectively. Let's take a deeper look at this mantra here.

Always be deciding

As the scope of your leadership responsibilities expands, the range of your decisions changes, and they tend to become more impactful. They are more about high-level strategy and finding the correct set of trade-offs and less about how to solve any specific engineering task.

"Always be deciding" highlights the critical role of decision making in leadership. Effective leaders understand the value of timely and well-informed decisions. They gather relevant information, consider the trade-offs involved, and make choices that align with the organization's goals and values. When faced with a difficult decision, you must be able to weigh the pros and cons of each option and make a decision that you believe is in the organization's best interests. Some trade-offs are obvious (e.g., diverting resources to the project, which adds more value to the organization), while others are ambiguous. Effective leaders would always make a timely decision in both cases.

Given any problem, Ben Collins-Sussman, a contributor to *Software Engineering at Google* (*https://oreil.ly/rBBcE*), suggests a three-step approach to decision making:

Identify the blinders

Blinders are mental blocks that prevent us from considering all possible solutions to a problem. They can be caused by biases, experiences, or the information available. Our view could be influenced by the experiences of people who have tried solving the problem before us. When we approach problem-solving with blinders, we limit our understanding of the problem and potential solutions. Blinders can lead us to overlook relevant information, disregard creative solutions, or fixate on a single approach. When you take over as a leader, you must first identify these blinders so that you can approach the problem with a fresh perspective and come up with a way forward in the form of a solution or additional questions that could help eliminate the blinders.

Identify the key trade-offs

Most problems at the leadership level will likely be nuanced and cannot be solved with a simple yes/no answer. You will often have to make trade-offs between different goals. For example, you may have to choose between a quick and easy solution to implement and one that is more effective but takes more time and effort. As a leader, it is your job to help your team make informed decisions about how to solve problems. This means calling out the trade-offs involved in different solutions, explaining these trade-offs to your team, and helping your team decide how to balance these trade-offs.

Decide, then iterate

Once your team has made a decision and acted upon it, you will need to evaluate the results and rebalance the trade-offs for the next round. This is the part where you are always deciding and responsible for continuously improving your engineering processes. You must explain to your teams when there is no perfect solution and where you all need to collectively iterate through different trials to find what will work best or be closest to the ideal solution. Waiting for the perfect solution can lead to a case of "analysis paralysis," resulting in indecision. The way forward in such a scenario is to make intentional decisions while being fully aware of the pros and cons.

A common decision that many tech team leaders face is whether to build something in-house or go for a commercial off-the-shelf solution for a specific problem and customize it. This could be a small utility library or an end-to-end product. Let's consider a decision faced by an engineering leader in charge of the development of a customer relationship management (CRM) system for the organization. The three-step approach to decision making recommends the following process in this scenario:

Identifying the blinders
: You may have a bias toward building a CRM system in-house because you believe it will give you more control over customization and integration with existing systems in the organization. However, this bias could lead to overlooking the time, resources, and expertise required for such a project.

Identifying the key trade-offs
: You need to analyze the trade-offs among building the CRM system in-house, outsourcing its development to a third-party vendor, and customizing an off-the-shelf CRM solution. You may consider factors such as cost, time to market, scalability, customization capabilities, and ongoing maintenance and support. While building something in-house would offer maximum control and customization, it would also require significant time, resources, and expertise. On the other hand, customization of an existing product would balance control and time to market, leveraging existing software with the flexibility to tailor it to specific needs. However, it may not fully meet all requirements, could require additional training for resources, and may cost more.

Deciding, then iterating
: After weighing the trade-offs, you may decide to initially customize an off-the-shelf CRM solution to meet your immediate needs while minimizing development time and costs. However, you remain open to revisiting this decision as the business grows and requirements evolve. You recognize that you may need to build certain functionalities in-house in the future to accommodate changing needs and priorities.

This example demonstrates how the three-step approach can guide your decision making in complex scenarios by helping you acknowledge biases, evaluate trade-offs, and remain adaptable to evolving circumstances.

Thus, identifying the blinders and key trade-offs and addressing them with a fresh perspective and informed decisions can help you overcome decision inertia for yourself and your teams. Deciding how to approach a problem is the first step to solving it.

Always be leaving

The second element of the always mantra, "always be leaving," may sound like you are abandoning your team, but it means that your team no longer depends on you for every decision. You have coached your team to self-sufficiency, and you can safely let go of the reins to move on to the next challenge.

Your team is genuinely self-sufficient when it can solve problems without you present. This implies that the team's core is so strong that engineering processes do not collapse when you are unavailable. Googlers use the "hit by a bus" metaphor to talk about people who are absent or leave the team or organization. The term "bus factor" denotes the number of people who need to get hypothetically hit by a bus before your project is ultimately doomed. A high bus factor means the organization is vulnerable to disruptions caused by losing key personnel. A low bus factor, on the other hand, means that the organization is more resilient to disturbances.

As a leader, you need to ensure that not only is the bus factor for your organization low but that you don't become the single point of failure (SPOF). If your team cannot progress in your absence, then you are an SPOF for the team. This is the exact situation that the "always be leaving" mantra teaches you to avoid. You want your team to be a self-driving machine. Let us see how you can build this machine:

1. *Divide the problem space*

   If you are leading a team of teams or a large project with various aspects, you should divide your problem space into several subproblems. For example, let's say your organization is building the proof of concept (POC) for a product that interfaces with multiple different systems for data and responds to user natural-language queries about the domain with relevant charts. While the various components of this project need to work together, you can loosely divide your problem space into modules for different interfaces, frontend design for charts, a language model for processing user queries, etc. Note that a loose structure allows you to restructure your subteams occasionally.

2. *Delegate subproblems to future leaders*

   Delegation is an essential tool when building your self-driving machine. Delegation frees up your time for more critical tasks and allows you to train a strong team of leaders. There may be some tasks that you do the best. Even so, ask yourself if another person can do the same work. There may be an initial learning curve, but gradually, you would end up creating a copy of your knowledge in another human being so that you are no longer the SPOF for that task. Conversely, there will be tasks that only you can do. Delegation helps you focus on such tasks.

3. *Adjust and iterate*

   Once you have your self-sufficient team, you can afford to distance yourself so that you are only macro-managing them. Direct your machine and keep it healthy with a subtle touch. This also creates opportunities for the leaders you identified in the previous step to grow in their careers. At the same time, you can move on to another similar problem space or an entirely different part of the organization. In short, you are now free to always be leaving.

Thus, in three steps, you could build a self-sufficient team and let go to ask yourself, "What next?" By embracing the "always be leaving" mindset, you can create space for new opportunities and allocate resources more effectively.

Always be scaling

The third mantra, "always be scaling," is not about the pursuit of growth. Instead, it's about protecting your most precious resources—*the limited pool of time, attention, and energy*—as you scale and grow aggressively. Failure to do so can affect your progress. You have to effectively scale yourself to handle additional responsibilities as your organization scales to solve various problems.

When you and your team successfully solve a problem, you receive more work and increased responsibilities besides accolades and appreciation. However, the quantity of precious resources remains the same. You cannot hire and delegate at the same pace as you scale. The key here is compressing your problems to make them smaller so that you can take on more significant issues. While you cannot really compress a problem, you can take steps to weaken its impact:

Force yourself to be proactive

In senior leadership roles, problems tend to arrive at your doorstep at their worst when everything else has failed. You are bombarded with escalations, all of which may be urgent. This can put you in a reactive mode so that you are forced to devote your time to issues that may be urgent but were not deemed important at first glance. If you had proactively worked out a strategy for dealing with a significant problem, it might not have come to you as an urgent issue. As a leader, you must focus on what's essential by dedicating a few hours a day to work on important but nonurgent tasks or delegating some of the urgent tasks.

Embrace the cycle of struggle and success

No matter how much you try, you will probably not be able to finish everything to your satisfaction. Rather than succumb to this struggle, learn to embrace it such that you are deciding what you can let go of. You can also apply Marie Kondo's philosophy (*https://oreil.ly/94DcF*) of decluttering your physical possessions to tackle your workload. If you divide your to-do list into three parts, you will probably notice these things:

- The bottom 20% of things are neither urgent nor important and easy to delete or ignore.
- The middle 60% might contain some bits of urgency or importance, but it's a mixed bag.
- The top 20% of things are absolutely, critically important.

Embracing the cycle of struggle and success implies allowing yourself to identify and focus exclusively on the top 20% of your list of critically important work that only you can do and ignore the rest.

Manage your energy

Besides protecting your time and attention, you must protect your energy levels to keep going in the long run. Being mindful of what gives you energy and what drains it would help. It's important to identify activities, tasks, and situations that energize and motivate you and those that deplete your energy. By recognizing these highs and lows, you can proactively prioritize activities that fuel your enthusiasm and delegate or minimize tasks that drain it. Additionally, prioritize self-care and recharge regularly. This includes taking real vacations where you can completely disconnect from work and rejuvenate. Plan dedicated weekends for relaxation and personal activities and take periodic breaks throughout the workday to refresh your mind. It's also essential to prioritize mental health and not hesitate to take a mental health day when needed.

If you think carefully, you might notice that the successful and effective leaders in your vicinity have mastered the principle of always be deciding, always be scaling, always be leaving. A prominent example that comes to my mind is that of leadership at Amazon.

Jeff Bezos, the founder of Amazon, famously said, "Day 2 companies make high-quality decisions, but they make high-quality decisions slowly. To keep the energy and dynamism of Day 1, you have to somehow make high-quality, high-velocity decisions." High-velocity decisions help Amazon experiment, iterate, and innovate quickly. Leaders are encouraged to disagree and commit. They must respectfully challenge decisions when they disagree. They must calculate the risk of their decisions. However, once a decision is reached, in spite of the risks involved, the participants commit wholly. With high-velocity decisions, leaders are always deciding.

Amazon's relentless focus on scale is legendary. Bezos's vision of building an "everything store"—a store that would sell nearly every type of product, all over the world—has translated into aggressive infrastructure expansion and automation over the years. Amazon has been constantly scaling and pushing the boundaries of how large and efficient a tech company can be.

Another famous Amazon principle is the two-pizza rule—a simple but effective management principle that involves limiting the size of a team to the number of people who can be fed by two pizzas. By keeping teams small and agile, Amazon is able to move quickly and respond to customer needs in real time. The two-pizza rule fosters a dynamic environment where people can readily move between teams and projects based on evolving needs (i.e., they can be constantly leaving).

The mantra of always be deciding, leaving, and scaling starts at a point after you have already tasted success. It gives important pointers on how to continue climbing the success ladder in the face of growing complexity, increasing numbers of projects, and more responsibilities. It tells us that to lead effectively when our workload doubles, we need not work twice as hard. By following this mantra, you can continue to expand your effectiveness model across all your teams.

Conclusion

The 3 E's model of effectiveness is designed to help you, your team, and even your organization as a whole keep pace with change. It guides you through the different stages of growth, and it starts by having you define what effectiveness means to your business domain and how you can initialize it within your teams. It tells you how to strengthen team effectiveness using known paradigms and models at the individual and team levels. Finally, knowing that one of these models is going to work for you, the 3 E's model prescribes proactive steps that you can take to help you continue being effective as your domain expands.

This concludes your introduction to effectiveness for software engineering teams. I have explained what makes software engineering teams effective; how one can distinguish and measure effectiveness, productivity, and efficiency; the importance of focusing on outcomes versus outputs; and implementing sustained effectiveness in a growing organization using the 3 E's model. In the second part of this book, I will cover specific areas to support everything I have shared so far. See you on the other side.

Effective Management Behaviors: Research from Google

In Chapter 1, we looked at how Google's Project Oxygen and Project Aristotle identified key factors that make teams effective. In this chapter, we'll dive deeper into these research projects, but from the perspective of how their findings reveal what makes a good engineering manager.

Examining this research helps us understand the specific behaviors and practices of effective engineering leadership. We can then adopt and develop these behaviors ourselves to lead high-performing teams in organizations of any size.

In this chapter, I will walk you through the list of essential behaviors for engineering managers as identified by Project Oxygen. I will discuss each and how you could develop these behaviors yourself. Similarly, I will also discuss what the findings of Project Aristotle regarding factors that contribute to team effectiveness could mean for you as a manager.

Project Oxygen

Project Oxygen was a research initiative launched by Google in 2008 to identify the qualities of a great manager at Google. The project was led by the People Analytics team—a team of researchers that studies employee data and finds ways to create and maintain a great workplace at Google. Before going into the details of this project, let's get the story of what Google was like before the project was launched.

A BRIEF HISTORY

Historically, some highly motivated engineering teams questioned the need for managers. They believed they could manage their own work without oversight. In its early days, Google even experimented with a completely flat organizational structure that eliminated middle management. However, as the company grew, it became clear that good managers were critical to team performance.

Fast-forward to 2007, when Google pioneered setting up a People Analytics team as part of its human resource function to help tackle employee well-being and productivity issues.

The team started examining the effect managers have on their teams. It worked to prove that "managers didn't matter." It used performance reviews and exit interview data to determine if better management helped reduce turnover or improve employee satisfaction. The data revealed that teams with great managers were, in fact, happier and more productive. Managerial responsibilities go beyond overseeing day-to-day work. They must also support employees' personal needs, development, and career planning. Once it had established that employee well-being and productivity were related to manager quality—that managers *do* matter—the team launched Project Oxygen to determine *what makes a manager great at Google.*

RESEARCH PROCESS

To answer the question, "*What makes managers great at Google?*" the People Analytics research team conducted detailed employee surveys targeting manager quality. It asked employees about their managers and, at the same time, conducted double-anonymized qualitative interviews with managers, asking questions such as "How often do you have career development discussions with your direct reports?" and "What do you do to develop a vision for your team?"

The team also identified the best and worst managers at Google, based on comments from surveys, performance reviews, and submissions for the company's Great Manager Award. Managers from all levels and geographies, and from Google's three primary functions (engineering, global business, and general and administrative), participated in the study.

Combining qualitative interview data and knowledge of the best and worst managers revealed illustrative examples of effective and ineffective management styles.

The research helped Google identify 8 key behaviors common to high-performing managers at Google. Over the years, these were incorporated into Google's manager development programs and further updated to 10 key behaviors. I will discuss these behaviors in detail in the next section.

BEHAVIORS OF HIGH-PERFORMING MANAGERS

Project Oxygen's expansion of the behaviors of high-performing managers from 8 to 10 goes to show that people are always learning. These behaviors are taught at Google manager training but can be used at any organization.

The 10 key behaviors are summarized in Figure 4-1.

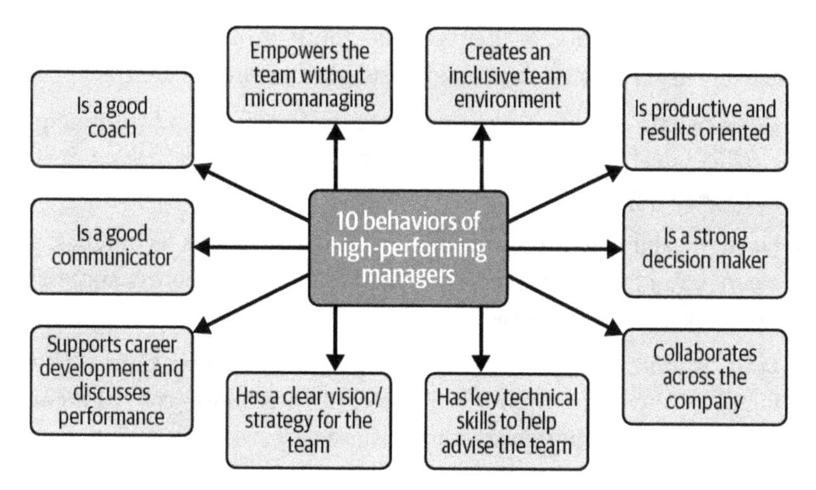

Figure 4-1. The 10 behaviors of high-performing managers

You'll notice some of these behaviors, such as coaching and providing feedback, have surfaced in earlier chapters in the context of effective leadership. All of these behaviors are interconnected aspects of being a great manager that enables team success. Let's examine each one in more detail.

Is a good coach

Great managers coach their team members to develop their skills and reach their full potential. Good coaches are patient, empathetic, and supportive. They are self-aware, good listeners, and able to give constructive feedback. They don't act like coaches only when it's convenient or when they need something from their coachees; they genuinely care about the coachee's success.

To become a good coach, you should do the following:

- Give thoughtful, specific, well-modulated, and constructive feedback on a timely basis.
- Offer guidance and articulate your expectations, ensuring that others understand and can implement the message.
- Hold regular one-on-ones with direct reports to coach them on their performance and discuss career goals.
- Tailor coaching to each individual's strengths, motivations, and development areas.
- Ask good questions to help people think through their options. Also actively listen to answers and understand where others are coming from.
- Demonstrate empathy to create a safe and supportive environment where coachees feel comfortable expressing themselves, discussing challenges, and seeking guidance.
- Have the ability to motivate and inspire using various techniques, such as positive reinforcement, setting high standards, celebrating successes, and fostering a growth mindset.
- Lead by example and embody the qualities and behaviors you want to instill in your team. Demonstrate integrity, resilience, commitment, and a growth mindset through your actions.

Empowers team without micromanaging

A good manager empowers their team members to do excellent work by giving them space to make decisions and solve problems independently. Empowerment without micromanagement is key. Here's how you can empower your team:

Offer stretch assignments
: Ask team members if they have the bandwidth to work on something that's beyond their current knowledge or skill level. Such assignments (also known as stretch assignments (*https://oreil.ly/5Z49b*)) give them an opportunity to learn, grow, and "stretch" themselves developmentally by placing them in a challenging position.

Intervene judiciously and with a light touch
: Balance the autonomy you offer team members with being available and dependable when someone needs advice. For example, after assigning a task that should take five days, I don't ask my team to give me a daily update, but I would expect it to let me know immediately if it is not able to proceed for some reason.

Encourage autonomy
: Allow team members to identify solutions independently, standing behind to steer them in the right direction only when needed. I once heard a Googler say these words of appreciation for their manager: "Although he gives me the freedom to work toward my goals, he knows when to intervene and advises me not to pursue problems unnecessarily if things don't go well."

Encourage innovation and thoughtful risk-taking
: Encourage your team members to think out of the box by allowing them some time to further analyze or try out their ideas. Reward successful initiatives to build trust.

Be a cheerleader for your team
: Advocate for your team to those outside it, be they senior management, another team, or end users. Communicate your team's achievements and ensure that they get the recognition they deserve. Give credit where it is due; this can be an essential aspect of motivating your employees (and yourself).

Provide constructive feedback
: Provide feedback where helpful in a way that helps employees grow professionally—not only by discussing achievements but also by helping with areas where there's room for improvement.

Empowering team members and providing autonomy means giving people space within which they can grow as individuals and professionals. A great manager understands that this is one of the best ways for them or anyone else on their team to develop professionally—and that such growth should never stop!

Creates an inclusive team environment, showing concern for success and well-being

A good manager understands that to create an inclusive team environment, one has to show concern for success and well-being. They create a safe space for team members to share their ideas and opinions, encourage them to think creatively and solve problems, show concern for the health of colleagues by offering support if they're having personal issues, encourage team members to learn from mistakes without judgment, and take time to get to know each team member personally.

Some of the things that you can do regularly to create an inclusive team environment include the following:

Make new team members feel welcomed
: Proactively welcome new team members and help them integrate into the team. You can assign a buddy or mentor to guide the newcomer, introduce them to team members, and provide necessary resources and information to facilitate a smooth onboarding process. This helps new team members feel valued and part of the team from the start.

Build rapport within the team
: Encourage team-building activities and facilitate opportunities for team members to connect and bond with each other, helping build rapport within the team. Team lunches, social events, and team-building exercises provide avenues for fostering positive relationships and open communication channels.

Be an enthusiastic cheerleader to support the team
: Managers play a crucial role in motivating and supporting their team members. Celebrate team successes, acknowledge individual achievements, and provide recognition and positive feedback. By being an enthusiastic cheerleader, you can boost morale, instill confidence, and create a positive and uplifting environment where team members feel valued and motivated to excel.

Role-model civility
: Leading by example, managers should demonstrate professionalism, treat team members with dignity, and promote inclusivity. When managers exemplify respect for diverse perspectives, the team will follow suit in fostering an environment where everyone feels welcomed and valued.

Actively care about, understand, and support team members' well-being
: Actively caring about team members' well-being involves regular check-ins with individuals to ask about their workload and provide resources or support when needed. By actively listening, showing empathy, and addressing concerns, you can create an environment where team members feel supported.

Show support in good and bad times
: Being a cheerleader also involves supporting your team during difficult times, such as setbacks or failures. At such times, you can offer encouragement, provide assistance, and foster a culture of learning from mistakes rather than blame. By demonstrating unwavering support, you can cultivate a sense of trust and loyalty within the team.

Create psychological safety on the team
: This involves encouraging open dialogue, actively listening to different perspectives, and addressing conflicts or issues promptly and respectfully. When team members feel psychologically safe (as described in Chapter 1), they are more likely to contribute their unique insights, fostering innovation and collaboration.

A good manager is willing to learn from their team members and collaborate with them. A good manager is flexible enough to change their approach if one isn't working well, and they only force their ideas on others if necessary.

Managers at Google are encouraged to show concern for employees' well-being using a technique called One Simple Thing (*https://oreil.ly/a56hr*). The idea is to let employees identify *one simple thing* that could help improve their well-being and work-life balance. The goals are non-work-related and easy to pursue. For example, "I will take a one-hour break three times a week to work out," or "I will disconnect on a one-week vacation this quarter." You can start the same practice in your team, too.

Is productive and results oriented

You can be productive and get results, but you must be results oriented, working toward set goals to manage effectively. This is not just about setting the bar high but also about pushing for outcomes. It's about working with your team to understand what success looks like and how you can get there together. It means setting clear goals and objectives, having regular meetings to track progress against them, and making adjustments as needed. It also means holding yourself accountable for the results of your team.

The following are ways to help drive your team toward achieving goals and delivering successful outcomes:

Assemble a diverse team
: A diverse team includes people from different cultural backgrounds and with varied skills, experience levels, and perspectives. Results-oriented managers understand the value of diversity. You can foster creativity, innovation, and problem-solving by bringing together a diverse range of talents and viewpoints, leading to more robust and well-rounded outcomes.

Translate the vision/strategy into measurable goals
: Effective managers translate the overarching vision or strategy into clear, measurable goals for their teams. Break down larger objectives into smaller, actionable milestones that your team members can understand and work toward. By setting specific and measurable goals, you provide a clear direction and focus for your team, ensuring that efforts align with the overall vision.

Structure the team and allocate resources to achieve their goals
: Results-oriented managers understand the importance of structuring their teams to optimize efficiency and productivity. Assess the skills and expertise of team members and allocate tasks and responsibilities accordingly. Additionally, ensure that the team has the necessary resources to accomplish its goals effectively, such as budget, technology, and training.

Be clear about who owns what
: Goal-driven managers establish clear lines of ownership and accountability within the team. Define roles and responsibilities, ensuring that each team member understands their specific areas of ownership. This clarity prevents confusion and duplication of efforts and ensures everyone is aware of their individual contributions to the team's overall success.

Remove any roadblocks the team may have
: Goal-driven managers actively identify and remove any roadblocks or obstacles that may hinder the team's progress. You should proactively address challenges arising from internal processes, resource constraints, or external factors. By facilitating problem-solving, providing guidance,

and leveraging their influence, goal-driven managers enable the team to overcome obstacles and stay on track toward achieving its goals.

Be foresighted and plan for potential risks

When you are focused on results, you tend to have the foresight to anticipate changes and plan for them so that you can guide your team through these changes toward a successful outcome. To prepare for challenges, you can stay informed about industry trends, market shifts, or internal organizational changes and proactively plan for these potential disruptions. By having contingency plans and adapting strategies in response to changes, you ensure that your team remains agile and can navigate uncertainties toward successful outcomes.

Is a good communicator—listens and shares information

Communication is crucial for any discipline, and management is far from being an exception. Managers communicate with their teams in many ways, such as email, messenger, meetings, tools, etc. I will discuss effective techniques for using these in detail in Chapter 6. In general, though, as a good communicator you should do the following:

- Encourage open discussion.
- Always aim to be responsive.
- Share information from leaders and explain the context.
- Be honest, even when the truth is unpleasant.
- Be calm under pressure.
- Listen to other team members.

In addition to being good communicators, good managers are also willing to share information with their team members. They make sure that everyone on the team has an opportunity to contribute and participate in any important decisions.

Listen carefully when your team members speak with you. While you may not necessarily agree with everything they say, understand that it is vital for your team members to have someone who will listen to them without judgment or bias.

A one-on-one meeting with each of your team members is an excellent opportunity to connect directly with team members and give them individualized attention. To streamline your meeting, consider using a templated agenda that ensures that you have a fulfilling discussion.

Effective One-on-One Template

The following is an example outline for an effective one-on-one meeting:

Manager:

- Vacation—how did the team's restaurant suggestion work out? [Showing you care]
- Share kudos from the director's staff meeting on Project X's impact. [Big picture]
- What have you been up to? [Checking in/catching up]
- What can I help you with? [Removing roadblocks/obstacles]
- What else? [Expansive]
- Is there anything I should be doing for you that I'm not doing? [Checking for your effectiveness]

Team member:

- What I did last week: update on Project Y
- What I plan to do this week: deliver design document for v2.0 of Project X
- Flag possible delay on subproject of Project Y
- Follow up on ABC discussion from last week's one-on-one
- Discuss interest in doing project with Team C

While regular one-on-ones often include project updates, be sure to also cover bigger-picture topics like career growth, motivation, and general well-being. The goal is to connect individually with each team member and understand how to best support them.

Good managers aren't afraid of asking questions, either; if something is unclear or confusing, ask for clarification before making assumptions about what was said or how it was meant. You can admit when you don't know something rather than pretending you do (which may lead to poor decisions). Of course, this doesn't mean that you shouldn't try your best at everything—focus on getting better at things over time instead of pretending nothing happened!

Supports career development and discusses performance

Good managers are honest and open about their expectations, giving the team members plenty of feedback so that they can improve. They also provide constructive criticism in a way that encourages team members to grow rather than discouraging them from trying at all. They don't wait until performance reviews to give feedback but give it regularly throughout the year. Good managers also give their team members opportunities to grow. They help them learn new skills and provide the resources they need to succeed in their careers.

To become a good career developer, you should do the following:

Communicate performance expectations
: Clearly communicate performance expectations to your employees. Establish clear goals, milestones, and performance standards, ensuring that employees understand what is expected of them. By providing this clarity, you enable employees to align their efforts and performance with organizational objectives.

Give employees fair performance evaluations
: Conduct fair and objective performance evaluations. Provide timely and constructive feedback to employees, highlighting their strengths, areas for improvement, and potential growth opportunities. By offering honest assessments, you facilitate the professional growth of employees and enable them to make informed decisions regarding their career paths.

Explain how compensation is tied to performance
: Employees need to understand how their compensation is tied to their performance. Explain the performance-based compensation structure, including the criteria used to determine salary increases, bonuses, or other rewards. By clarifying this connection, you can provide a transparent and equitable compensation system.

Advise employees on career prospects other than promotion
: Career growth does not always mean climbing the hierarchical ladder. Advise employees on alternative career paths within the organization, such as lateral moves, special projects, or cross-functional opportunities. By broadening employees' understanding of career prospects, you can help them explore diverse avenues for development and advancement.

Help team members find ways to grow and change within the company
: Actively assist your team members in identifying and pursuing growth opportunities within the company. Encourage employees to explore professional development programs, training initiatives, certifications, or stretch assignments that align with their interests and career aspirations. By doing this, you foster a culture of continuous learning and development by supporting employees in their growth journeys.

Google uses the GROW (Goal, Reality, Options, Will) model to structure career development conversations between managers and their direct reports. The model is based on four key questions:

Goal
: What do you want? This question assesses a person's career aspirations, dream role, motivation, and values.

Reality
: What's happening now? This question tries to gauge what a person feels about their current role—if they feel challenged or frustrated.

Options
: What could you do? Discuss options that could help take a person from the current state (reality) closer to achieving their goals.

Will
: What will you do now? Identify the best option and steps to help someone start on that path.

You can use a similar structure when conducting performance reviews or career progression discussions with your direct reports.

Has a clear vision/strategy for the team

Good managers know their team needs a clear vision and strategy that is communicated effectively. They help their team understand the "why" behind their actions. Good managers lead by example and can motivate their team members through their passion for the company's mission and vision. They hold regular meetings and one-on-ones with team members to ensure everyone is on the same page regarding goals, plans, and priorities.

Managers with a clear vision can inspire and align their team members toward achieving organizational goals. Here's how you can do it, too:

Create a vision/strategy to inspire team members
: Craft a compelling narrative that inspires and motivates team members. Develop a vision that paints a vivid picture of the team's future success, highlighting the impact and value of its work. Thus, you can ignite enthusiasm and commitment to shared objectives.

Align the team's vision/strategy with the company's
: Ensure that you understand the company's mission, values, and strategic objectives, and then translate those into a vision that aligns with the team's purpose. This creates a sense of purpose and direction that drives collective effort.

Involve the team in creating the vision where it makes sense
: Wherever possible, solicit input, ideas, and perspectives from team members, fostering a sense of ownership and empowerment. By involving the team in the decision-making process, you can tap into diverse expertise, promote collaboration, and create a shared vision that reflects the team's collective wisdom.

Clearly communicate the vision, helping the team understand and encouraging questions
: Use clear and concise language, avoid jargon, and ensure everyone easily understands the message. Provide context, explain the rationale behind the vision, and encourage team members to ask questions and seek clarification.

Help the team understand how the overall strategy translates to its work
: Learn how to bridge the gap between the overall strategy and the day-to-day work of your team members. Help team members understand how their contributions fit into the larger strategic framework. You can enhance employee engagement, motivation, and a sense of purpose by explaining the direct impact of the team's work on the achievement of strategic goals.

Google recommends using the following steps, shown in Figure 4-2, to help teams define their values and connect them to their short-term goals:

Core values
: The team's deeply held beliefs

Purpose
: The reason why the team exists

Mission
: What the team is trying to achieve

Strategy
: How the team plans to realize the mission

Goals
: Short-term, achievable objectives to implement the strategy

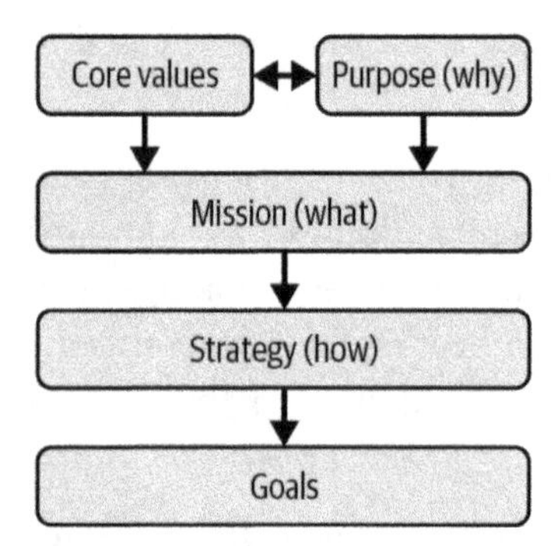

Figure 4-2. From values to goals

Your core values and purpose help you define your mission. Once you know your mission, you can develop a long-term strategy to achieve it. This then helps you identify the short-term goals you need to achieve at each stage to ensure the successful implementation of your strategy.

Has key technical skills to help advise the team

Managers can use their technical knowledge and context to advise their team in the following ways:

Help the team navigate technical complexity
: You should be able to understand the technical complexities that your team faces. You are familiar with the overall intricacies of the projects and can guide your team members through challenges and obstacles. You are able to provide clarity, direction, and support, helping the team navigate complex technical issues and find practical solutions.

Understand the challenges of the work
: Managers of technical teams should have firsthand knowledge and experience in the technical domains their teams operate in. You should understand the challenges and intricacies of your team members' work, including the tools, processes, and methodologies involved. This understanding will help you to empathize with your team, anticipate potential roadblocks, and provide relevant guidance and support.

Use technical skills to help solve problems
: You should be able to leverage your expertise to contribute to problem-solving efforts. You can actively engage with the team in troubleshooting and finding innovative solutions. By sharing your technical knowledge and insights, you provide valuable input that helps the team overcome technical hurdles and achieve project objectives.

Learn new skills to meet business needs
: You should recognize that staying updated with the latest trends, tools, and technologies in a dynamic landscape is vital. You can invest in learning and expanding your technical skills to meet evolving business needs. By keeping yourself abreast of industry advancements, you can make informed decisions, provide relevant guidance to your team, and ensure that technical strategies align with organizational goals.

Bridge the gap between technical and nontechnical stakeholders
: Develop strong communication skills such that you can effectively bridge the gap between technical and nontechnical stakeholders. This will help you translate complex technical concepts into easily understandable terms for stakeholders who may not have a technical background. Thus, you can

ensure alignment between technical solutions and business objectives by facilitating clear communication and understanding.

Is a strong decision maker

One of the most essential traits for a manager is being able to make vital decisions. This is especially important for managers of technical teams since they are often called upon to make decisions about the direction of their team and company. They need to be able to weigh different options and select the best one based on their knowledge of technology, business needs, and goals.

The best managers can look at a problem from all angles and make decisions based on the information they have. They can avoid getting stuck in their way of thinking, which is essential when dealing with complex problems requiring many different viewpoints.

You can become a strong decision maker using the following strategies:

Make decisions efficiently, with the best interests of the business in mind
: Demonstrate a sense of urgency when needed, ensuring that decisions are made promptly without unnecessary delays. Approach decision making objectively, basing judgments on facts, data, and logical reasoning rather than personal biases or emotions.

Make decisions with an eye toward the situation at hand
: Consider the unique characteristics and context of each situation before making a decision. Analyze the specific circumstances, including the nature of the problem, the available resources, the potential risks, and the desired outcomes. Adapt a decision-making approach based on the situation's complexity, urgency, and potential impact.

Clearly communicate decision-making rationale
: You should be able to communicate your rationale clearly and transparently. You will often need to justify the decision by explaining the factors considered, the analysis performed, and the reasoning behind the decision. By communicating the rationale, you can help others understand the decision's logic and align their understanding with the desired outcomes. This transparency fosters trust, reduces ambiguity, and enables stakeholders to effectively support and implement the decision.

Collaborates across the company

The best managers collaborate across the business to ensure that their teams work together effectively. They don't just focus on their departments or teams; they take a more holistic view of the company. They understand that cooperation between different groups is essential to the success of any project, and they do what they can to facilitate those interactions.

To become a good cross-functional collaborator, you should consistently do the following things:

Prioritize collective goals and outcomes that align with the overall business objectives, even when they require cross-functional collaboration

This means proactively seeking opportunities to partner with other teams where it will produce the best results for the company, rather than focusing solely on one's own team's deliverables. It involves fostering a mindset of "we're all in this together" and making decisions based on promoting the collective good.

Seek opportunities to partner with other teams where it will produce the best results for the business as a whole

Rather than waiting for cross-functional collaboration to be requested or mandated, proactively seek out chances to align your team's efforts with the company's overarching goals and strategies.

Role-model collaboration across different teams and functions

Lead by example and actively promote collaboration across different teams and functions. Encourage open communication, information sharing, and knowledge exchange. Proactively build relationships with stakeholders from various departments and work collaboratively to achieve shared goals. By demonstrating the value of collaboration, you can inspire others to do the same and contribute to a culture of cross-functional teamwork.

Hold your team accountable for following company practices/policies

Ensure that your team adheres to the company's practices and policies. Set clear expectations and hold your team members accountable for following established guidelines and protocols. By ensuring consistency and compliance, you can establish a harmonious working environment and demonstrate your commitment to the organization's values and standards.

Take part in the company's culture and community

Actively engage in the company's culture and community by participating in company-wide events, initiatives, and social activities. You can contribute to a positive and inclusive work environment by fostering a sense of belonging and promoting collaboration across different teams. Try to make the workplace welcoming for everyone, embracing diversity and creating an environment where all employees feel valued and included.

OUTCOMES

Project Oxygen accomplished what it set out to do and proved that managers mattered. It then took it a step further and also quantitatively established and institutionalized the essential qualities of great managers. The team took the concept of data-driven continuous improvement and applied it successfully to the soft skills of leadership, communication, collaboration, and ultimately management.

The behaviors identified in the project have had a positive impact on Google employee performance, satisfaction, and turnover. The project has also helped Google create a more effective and efficient management team. The findings have been incorporated into Google's manager development programs. These programs help new and existing managers learn the skills and behaviors necessary to be great managers at Google. In fact, Google revamped the selection criteria for their annual Great Manager Award to reflect the Project Oxygen behaviors.

Widespread adoption of Project Oxygen behaviors has significantly impacted how employees rate the degree of collaboration at Google, the transparency of performance evaluations, and their groups' commitment to innovation and risk-taking. These factors have contributed to the following:

- Increased employee performance
- Improved employee satisfaction
- Reduced employee turnover
- Better decision making
- Increased collaboration
- A more positive work environment

Insights from Project Oxygen findings are valuable for any organization that wants to improve its management team. The behaviors identified in the project are essential for any manager who wants to be successful. Having seen what it meant for Google, you can now see how a similar strategy can be implemented in other organizations irrespective of size and type.

LEVERAGING PROJECT OXYGEN'S FINDINGS

With its focus on innovation and a firm commitment to user experience as well as employee well-being, Google has been a unique organization since its birth. It is also a large organization that practices diversity in product offerings and people culture. While the insights from Project Oxygen are valuable and helped make a significant difference to work life at Google, how can it help you as a software engineering manager?

There are two key aspects to this conversation:

Your current background and skill set

You can self-reflect on your background and current skill set in relation to Project Oxygen's findings. One way of doing this is using the findings as a checklist and identifying areas where you need to meet the requirements or fall short. For example:

- Perhaps you recognize that you need more experience in effective coaching. You can then proactively seek opportunities to develop your coaching skills. Identify someone on your team who may benefit from your guidance. Create a plan of action tailored to their needs. Discuss with them and let them know you welcome feedback on how to improve as a coach.
- Maybe you need to improve at conducting one-on-ones. You can then use a template for structuring one-on-one meetings and tailor it through trial and error to improve your skills. Informal feedback from team members may prove helpful in this scenario.

You can also proactively address skill gaps by seeking training and development opportunities for managers. Find workshops, webinars, or courses explicitly focusing on the behaviors and qualities highlighted in Project Oxygen. This can be a strategic way to align personal development with the research findings.

Manager assessment practices at your organization

If your organization has established manager assessment practices, you can gain valuable insights into what your team thinks of you as a leader—your management style, strengths, and areas that require improvement. For instance, if feedback indicates a lack of clear communication, you can focus on improving this aspect.

Without formalized assessment, you can gain insights through discussion and active listening. Understand what your team is trying to tell you. Ask questions like "Are you happy with the opportunities given to you, or would you like additional stretch opportunities?" This should help you assess directions for improving as a manager. You can use Google's manager feedback form (*https://oreil.ly/xJ_hU*) for more such questions.

The findings of Project Oxygen and the tools developed by Google based on these findings are a solid source of reference material for organizations and teams that want to improve.

In the next section, let's look at another research project from Google: Project Aristotle.

Project Aristotle

In Chapter 1, we explored the findings of Project Aristotle, a research study that identified the key factors contributing to team effectiveness at Google. The study found that the five dynamics shown in Figure 4-3, were critical to building successful teams.

In that chapter, we examined these factors from the perspective of what makes a team effective and how to foster those attributes to build a high-performing team.

Now, let's shift our focus to how managers can foster these same dynamics in their teams. As a manager, you play a crucial role in creating an environment that supports and nurtures these key factors of effectiveness.

Figure 4-3. Project Aristotle key dynamics

PSYCHOLOGICAL SAFETY

Recall from Chapter 1 that *psychological safety* refers to a shared belief that the team is safe for interpersonal risk-taking. It is the most critical factor in team effectiveness, as shown in the Project Aristotle research.

The idea of team psychological safety was first proposed by Amy Edmondson in her paper "Psychological Safety and Learning Behavior in Work Teams" (*https://oreil.ly/WXzka*). Edmondson defines team psychological safety as a "shared belief that the team is safe for interpersonal risk taking." She also talks about how this is different from "group cohesiveness." *Group cohesiveness* is about getting along with each other as a group, while team psychological safety is about feeling safe to speak up and take risks in a team without being afraid of looking bad or hurting others' feelings. It's about getting along while being able to have open and honest discussions where everyone feels respected and trusted.

Cohesiveness in a team can reduce the willingness to disagree with fellow team members or to challenge others' views. Psychological safety suggests that team members are neither too careless nor too optimistic when interacting with others on their team. This creates an environment where team members don't fear embarrassment or outright rejection when speaking up; they feel comfortable asking even basic questions without worrying about being perceived as ignorant. There is a sense of mutual respect and trust among team members, allowing each person to be themselves.

Note

While psychological safety does not play a direct role in fulfilling customer requirements, it ensures that a team as a whole will take appropriate actions to accomplish its goals.

Some may argue that a few employees not speaking up could be just because they are too shy or introverted to stand up and articulate their views in front of a team. A study (*https://oreil.ly/1smQm*) published in the *Harvard Business Review* shows that while indeed this personality perspective could be one of the reasons, the other reason, called the situational perspective, is more predominant. The *situational perspective* is where employees fail to speak up because they feel their work environment is not conducive to it (*https://oreil.ly/6NPbA*). The study found that strong environmental norms could override the influence of personality on employees' willingness to speak up at work.

For her study on the impact of team psychological safety on organizational learning and performance, Edmondson asked participants whether they agreed or disagreed with statements such as "If you make a mistake on this team, it is often held against you," and "No one on this team would deliberately act in a way that undermines my efforts." As an example of a practical application of this research, Google held workshops where they role-played scenarios to illustrate behaviors that can support and harm psychological safety. These workshops helped people understand that we usually try to protect ourselves by refusing to behave in a way that could negatively influence how others perceive our competence, awareness, and positivity. However, this behavior is detrimental to effective teamwork. The safer team members feel with one another, the more likely they are to admit mistakes, partner, and take on new roles.

Managers who want to foster psychological safety in their teams must understand that doing so does not mean relaxing performance standards. You cannot allow inappropriate conduct just to ensure team members feel safe. Performance standards and psychological safety must both be high for people to speak up and voice concerns. Here are a few steps managers can take to promote psychological safety in their teams:

Approach conflict as a collaborator, not an adversary

Look at conflicts optimistically as opportunities for growth and learning. Conflicts need not turn into confrontations. Instead of taking sides or assigning blame, managers should facilitate open and respectful discussions where all parties involved can express their perspectives. They should encourage the team to focus on the shared goal of finding a resolution and promoting understanding.

Speak human to human

Effective communication is crucial for establishing psychological safety. Managers should strive to communicate with their team members in a relatable and empathetic manner. This means avoiding jargon or overly formal language and using clear, concise, and inclusive communication styles. By speaking human to human, managers can foster connections, build trust, and create an environment where individuals feel comfortable expressing their thoughts and concerns.

Anticipate reactions and plan countermoves

Managers should proactively anticipate potential reactions or resistance from team members, especially when discussing sensitive or challenging topics. By considering different perspectives and possible reactions in advance, managers can better prepare their responses and address concerns or fears in a supportive and understanding manner. This proactive approach demonstrates that managers are attentive and empathetic, contributing to a culture of psychological safety.

Replace blame with curiosity

Instead of assigning blame when mistakes occur or something goes wrong, managers should cultivate a culture of curiosity. They can encourage team members to explore the underlying reasons, contributing factors, and lessons learned from a situation. This shift in mindset from blame to curiosity helps create an environment where mistakes are seen as opportunities for growth and learning rather than reasons for punishment or shame.

Ask for feedback on delivery

Managers should actively seek feedback from their team members regarding their communication and leadership styles. By asking for feedback on how they deliver messages, managers demonstrate their openness to improvement and commitment to creating a psychologically safe environment. This feedback loop allows managers to understand how their communication may impact team members and to make adjustments accordingly.

Measure psychological safety
: It is essential for managers to regularly assess and measure the level of psychological safety within their teams. This can be done through anonymous surveys, focus groups, or one-on-one conversations. By collecting feedback, managers gain insights into areas that need improvement and can take targeted actions to enhance psychological safety. Regular measurement also demonstrates a commitment to continuous improvement and reinforces the importance of psychological safety as a team value.

Managers looking to reinforce psychological safety and interpersonal relations in their teams can also use or tailor Google's guide for managers (*https://oreil.ly/A_Xfo*).

DEPENDABILITY

For a team to accomplish more than an individual would, teammates should be able to trust each other and work together. Dependability within a team is high if the team members can count on each other in various respects. Dependability fosters trust. Team members should be able to rely on each other and their leaders to do the right thing.

Something as simple as being punctual to team meetings could imply dependability. One person not being on time can lead to frustration for the entire team. Moreover, others might become callous about punctuality if they think being late is OK.

While it's not easy to identify a dependable team member at first glance, every small action can contribute to your reputation as a reliable or unreliable person. Dependable team members and leaders demonstrate some essential qualities such as these:

Genuine intentions
: A dependable team member has a solid ethical foundation, because of which their intentions are genuine. They prioritize the team's collective goals over personal ambitions and demonstrate a commitment to acting in the best interests of the team and its members. They are driven by a sense of integrity, honesty, and fairness, which fosters trust and confidence among their colleagues.

Accountability

Accountability is a fundamental quality of a dependable team member. They take ownership of their assigned tasks and demonstrate responsibility for their actions and outcomes. They understand the importance of meeting deadlines, fulfilling commitments, and delivering high-quality work. A responsible team member takes initiative, proactively addresses challenges, and seeks solutions to problems, contributing to the team's overall success.

Sound thinking

Dependable team members demonstrate sound thinking involving critical and analytical reasoning. They approach problems and decision making with a logical and thoughtful mindset. They consider relevant information, evaluate different perspectives, and weigh the potential consequences before making informed choices. Sound thinking allows team members to contribute valuable insights and make reliable decisions that impact the team's objectives positively.

Consistent contribution

Dependable team members consistently contribute to the team's efforts. They actively participate in team discussions, meetings, and projects, offering their ideas and insights. They consistently deliver their work at a high standard and meet established deadlines. They are reliable and can be counted on to fulfill their commitments, strengthening the team's overall performance and trust in their capabilities.

As a manager, you are responsible for dependability within your team and the perceived dependability of your team for external stakeholders. In this regard, it's important to ask yourself the following questions:

- As a leader, do you set the tone for which behaviors are desirable within the team?
- Do other teams or internal and external stakeholders count on your team to meet commitments and deadlines?

Fostering dependability within the team can also ease the way to meeting deadlines and commitments. Even though it's an individual quality, you can take the following steps to foster dependability within your teams:

Lead by example
: Managers and team leaders should model dependability in their own actions and behaviors. Leaders set a standard for dependability that inspires the team to do the same by consistently meeting their commitments, following through on promises, and being transparent about challenges.

Promote collaboration and interdependence
: Emphasize the importance of teamwork and interdependence within the team. Encourage collaboration, where team members support and assist each other to meet shared goals. This fosters a sense of collective responsibility and reinforces the notion that dependability is a team effort.

Clearly define roles and expectations
: Clearly communicate individual roles, responsibilities, and performance expectations within the team. When team members clearly understand their tasks and what is expected of them, it promotes a sense of accountability and enables them to prioritize their work effectively.

Encourage open communication
: Foster an environment where team members feel comfortable discussing their workload, challenges, and potential bottlenecks. Encouraging open communication allows individuals to proactively address any issues or seek support when needed, ensuring that deadlines and commitments can be met.

Provide supportive feedback
: Give team members constructive feedback on their performance, highlighting their strengths and areas for improvement related to dependability. Encourage them to set personal goals for enhancing their dependability, and offer resources and guidance to support their development.

Dependability is vital to team effectiveness, as it builds trust, improves collaboration, and drives productivity. By clearly defining roles, encouraging open communication, providing supportive feedback, promoting collaboration, and leading by example, teams can foster a culture of dependability, leading to enhanced outcomes and success.

STRUCTURE AND CLARITY

One of the overall conclusions of Project Aristotle is that "at Google, who is on a team matters much less than how team members interact, structure their work, and view their contributions." Establishing a structure is essential for team members to understand what is expected of them. At the same time, clarity on goals and plans helps them work in the right direction. Team members should be clear about the overall structure and process and where they fit in it.

Clarity on structure and individual roles has become even more critical in the post-COVID world, which has seen a lot of volatility in organizations and where hybrid and remote work have become common. When engineers know who is working on which part of a project and what is expected from them, they can proceed independently and in the right direction without frequently checking in with others. On the other hand, ambiguity in structure and individual responsibilities can create stress and confusion.

Objectives and key results (OKRs)

In Chapter 2, I introduced how OKRs are a framework used by Google and many other organizations to establish clarity and alignment around goals. OKRs help create structure and process by defining what needs to be achieved and how success will be measured.

In Chapter 3, we touched on OKRs again in the context of communicating team goals and ensuring everyone understands how their work contributes to the bigger picture. Now, let's dive a bit deeper into how OKRs can be used to drive alignment between team and organizational objectives.

At Google, teams are encouraged to first align their OKRs with the company's overarching goals. While not every organizational OKR needs to be directly linked to a team OKR, there should be a clear connection between each team's objectives and at least one higher-level company objective. This ensures that the work of each team is contributing to the organization's most important priorities.

When setting their quarterly OKRs, Google teams start by identifying the key results that would have the greatest impact on advancing the relevant organizational OKRs. These become the team's top priorities for the quarter, with the rationale being that achieving these results will move the needle on the company's objectives.

Sample Team and Individual OKR

Objective: Accelerate [product] revenue growth

Key results:

Launch X feature to all users.

Implement X initiative to increase revenue per user by XX%.

Launch three revenue-specific experiments to learn what drives revenue growth.

Secure tech support to build XX feature in Q1.

Objective: Improve [product]'s reputation

Key results:

Reestablish [product]'s leadership by speaking at three industry events.

Identify and personally reach out to top XX users.

Shorten response time to user-flagged errors by XX%.

OKRs should be revisited and reviewed a few times per quarter. While this is not compulsory, OKRs can serve as a calibration tool, helping team members adjust to new information, abandon objectives that are clearly not achievable, and increase attention to borderline objectives that will benefit from additional resources. Having well-implemented OKRs can be an effective way to clarify goals at every level in the organization.

RACI matrix

Another popular framework that helps bring structure and clarity to assigning responsibilities to various roles is the responsibility assignment matrix, also known as the *RACI matrix* (*https://oreil.ly/VSJDJ*). It's a simple grid system that you can use to clarify people's responsibilities and ensure that everything the team needs to do is taken care of. The acronym *RACI* stands for:

Responsible
: People who do the work

Accountable
: The person who owns the work to be done

Consulted
: People who should review the work and give feedback

Informed
: People who need to be kept in the loop

When using this model to assign responsibilities, you need to identify the roles who will be responsible, accountable, consulted, or informed for different tasks or deliverables in the project. For example, Table 4-1 shows the RACI matrix for two deliverables.

Table 4-1. Sample RACI matrix

Deliverable	Responsible	Accountable	Consulted	Informed
Design document	Solution architect	Project manager	Team leaders	Developers
Code	Developers	Project manager	Team leaders	Testers

MEANING

Finding *meaning* at the workplace refers to experiencing a sense of purpose, fulfillment, and progress at work. *Meaning* can mean different things to different people. For an engineer, it could simply mean indulging a passion for coding or finding issues. It could also mean achieving a personal goal such as financial security, supporting family, or being part of something that can make a difference to society.

When team members find their work meaningful and experience a sense of purpose, they bring their best selves to work and use their greatest strengths to help the organization flourish. They feel connected to their work at a deeper level and are motivated to think creatively and share ideas. When they are driven by a purpose, a project or task becomes more than something that just needs to be completed: they want it to be a job well done. Team members are also more likely to readily embrace change if they believe that it is for the greater good of the team.

Purpose and *meaning* may seem like superficial terms at times. How do you even begin helping team members find their purpose? Motivational talks might fall flat. It's important to note that people find meaning at work when they know that their contributions matter. Helping people find meaning at work is about finding a relatable story that helps them realize that they love what they do.

As a manager, you can help them find their story and enhance their connection to their work. Every team member is different, and their motivations will likely differ. Your task is to help them customize their journey. Some things you can give team members to help them achieve this are as follows:

Positive feedback and recognition
: Employees want to feel appreciated and valued for their work. Managers can create a sense of purpose by providing regular feedback and recognition, both formal and informal. This can be done through one-on-one meetings, team meetings, and public recognition.

Strong support
: Creating a positive and supportive work environment helps employees feel valued and respected, leading to a stronger sense of purpose. Employees want to feel connected to their colleagues and their managers. Managers can help create strong relationships by being approachable and supportive, creating opportunities for team building, and celebrating successes together.

IMPACT

When people realize the true impact of their work, they're more engaged, innovative, and productive. Research (*https://oreil.ly/gjYQF*) shows that working on significant tasks can have a positive impact on job performance. When people know that their work matters, it creates momentum; people feel valued when their work is recognized. If no one notices their work, they might just stop caring for it and go lax in their execution of tasks.

Even when your team is a small part of a large organization, it is essential that your team members feel connected to the organization's goals through their work. They need to know that even the smallest tasks they work on help to shift an entire ecosystem forward. Part of your role as a leader is to help team members identify how they drive impact within and beyond the team.

As a manager, there are several steps you can take to help them understand why a project is essential and why they are a relevant part of the team. Even before joining an organization or a team, people may ask, "Why are you hiring me? What is expected from me? Why my role?" As a leader, you should be ready with an answer to these questions for every individual on your team at all times. Here are a few things you can highlight to help your team understand this better:

Connection to organization objectives
: Managers should go beyond simply assigning tasks and provide a clear connection between the organization's objectives and each team member's work. By explaining how their contributions align with the organization's broader purpose, managers help team members understand the significance of their roles. When employees recognize the impact their work has on achieving organizational goals, they feel a sense of purpose and are more motivated to perform at their best.

Working toward a team vision
: Managers can facilitate a collaborative process where your team co-creates a clear vision. Managers can foster a sense of ownership and commitment by involving team members in shaping the team's strategy with a focus on individual contribution. When team members have a shared vision, it becomes easier for them to see how their individual tasks contribute to the larger picture. This alignment promotes a sense of unity and collective effort, enhancing motivation and engagement.

Understanding the impact on clients and users
: Team members want to know that their work will make a positive difference in their clients' or end users' lives. If you cannot justify how a specific task enhances the experience or quality of a product or service, then it is likely that the team won't be so keen on doing it. Managers should regularly share success stories, customer feedback, or testimonials to highlight the real-world impact of their team's work.

Linking performance to outcomes
: Managers can establish a clear connection between individual and team performance and the outcomes they have achieved. OKRs help managers create a framework for tracking progress and success. Regularly reviewing and discussing performance metrics with team members helps them understand the direct impact of their efforts and fosters a results-driven mindset.

By employing these strategies, managers can help their teams gain a deeper understanding of the impact of their work. When team members have a clear sense of purpose, connection to organizational objectives, and awareness of the positive change they are making, they are more engaged, motivated, and invested in achieving shared goals.

OUTCOMES

Project Aristotle taught us that building human bonds rooted in empathy and conversational turn-taking was crucial to the psychological safety of the team and its success. Teams want to feel that their work is more than just labor. Project Aristotle encouraged conversations and discussions among people who might otherwise be uncomfortable sharing their feelings. Project Aristotle's results show us that in their bid to improve productivity and efficiency, leaders might end up missing cues about team members who feel excluded or uncomfortable expressing themselves.

To utilize the findings of Project Aristotle effectively, Google created a tool called the gTeams exercise, which I'll discuss in the next section.

Leveraging Project Aristotle's Findings

Just as Google created practices based on Project Oxygen to develop great managers, they also established a tool called *the gTeams exercise* to help teams strengthen the five dynamics uncovered by Project Aristotle.

The exercise starts with team members taking a 10-minute survey on how the team is doing in the five areas. The team then holds an in-person conversation to discuss the results and access resources to help it improve. Google teams that committed to regular practices to reinforce the dynamics, like starting meetings with casual chats to increase psychological safety, saw measurable improvement.

Over the years, thousands of Googlers across different teams have used this tool. Teams that adopted a new group norm—like kicking off every team meeting with informal chitchat about weekend plans—improved by 6% on psychological safety ratings and 10% on the structure and clarity ratings. A proven framework for team effectiveness has helped many teams improve their effectiveness.

While the gTeams exercise is a Google-specific tool, the concept is highly applicable to any team. Consider creating a simple survey for your own team to assess how you're doing on the five dynamics. Use the results to have an open discussion and identify practices to embed these factors into your team culture. Regular measurement will show you the impact over time.

With an understanding of Project Aristotle's findings, you now have a proven framework to evaluate and enhance your team's effectiveness, in addition to growing your own management skills.

Conclusion

Most of the findings of Project Oxygen and Project Aristotle may appear like common sense in retrospect. The significance of these projects lies in the extensive research conducted at Google and the data collected to support these findings. The research revealed the things that we should be doing deliberately, and the findings can more or less be universally applied in different workplaces. Although the research was done at Google, which is not a typical company in regard to global scale and resources, many findings can be adapted for your own practices, your own team, and your own organization.

Google's continued research has helped the company to understand the benefits of different practices that were implemented as a result of these projects and improvise as required. Your organization and team can also benefit by identifying which of these findings is relevant to you and tailoring your processes accordingly, one step at a time. Common sense or not, leaders must act on these findings and make them common practice within their teams and organizations to bring about a positive and measurable difference in their effectiveness.

In the next chapter, we will continue focusing on research, but this time of a different variety. So far, we have discussed the patterns you should implement for success; we will next talk about antipatterns that you should avoid to prevent failure.

5 Common Effectiveness Antipatterns

In the previous chapters, I discussed the importance of enabling effectiveness in engineering teams and detailed how factors such as psychological safety, dependability, and others foster effectiveness. I have also listed the steps managers, leaders, and engineers can take to become effective individually and as a team.

Some of you may have found this guidance intuitive. It's not as if one sets out to be ineffective! However, despite all your efforts, ineffectiveness will likely creep in when you least expect it to. You can blame the circumstances or Murphy's law, but there will be complications you couldn't have planned for, which can impact the project's outcome. Specific patterns in behavior and decision making enhance effectiveness, while others hinder your team's ability to achieve its goals. How can you enable the right patterns and protect your effective team from these pitfalls?

In software development, we have design patterns that are considered highly reliable and effective. In contrast, there are also antipatterns that are the exact opposite. *Antipatterns* are common responses to recurring problems that, although they may seem like solutions, are actually superficial, unproductive, and ineffective.

Rather than wait for the ball to drop, it's essential to study the typical antipatterns to effectiveness and identify them early when they occur. Having reviewed many pitfalls that tend to occur in software engineering teams, in this chapter, I will present the common antipatterns to effectiveness. In each case, you will learn the distinct characteristics that can help you detect them and tips to handle them to restore effectiveness in your team.

Recognizing patterns of behavior or practices helps team leaders proactively establish preventive measures. They can create guidelines, policies, or training programs to minimize the likelihood of antipatterns emerging in the first place. Identifying and addressing antipatterns can contribute to cultivating a healthy and productive engineering culture. This can help team leaders to introspect and identify areas where positive practices and collaboration may be needed.

Before I discuss the antipatterns that affect effectiveness in software engineering teams, I want to introduce a system for categorizing them.

Antipatterns Categorization

Antipatterns represent problem areas. The sources and causes of the various antipatterns that hamper team effectiveness can differ. Dysfunction can start anywhere, be it individual behavior, leadership practices, or team processes, but propagate across the team. The impact of antipatterns may not be apparent immediately. It may start with an individual team member but cause a slow burn to the entire team.

To better understand and address the different antipatterns, we must categorize them. Categorization aims to provide a clear and organized structure for grouping similar antipatterns together. As with software design patterns (*https://oreil.ly/zswdr*), it would be good to have a standard library of effectiveness antipatterns and their corresponding solutions. This would help team leaders identify and recognize recurring behavior patterns or structural issues that might hinder engineering processes.

Each category here represents a distinct type of antipattern based on where it originates. Understanding the category helps in choosing appropriate targets and strategies for mitigation. Team leaders can apply targeted solutions based on the type of antipattern to address the specific challenges faced by the team. Knowing the category can help you investigate the reasons behind certain behaviors, identify the root cause, understand the potential impact, and address the core issues more effectively.

Antipatterns can be grouped into the categories listed in Table 5-1.

Table 5-1. Antipattern categories

Antipattern	Description	Mitigation
Individual	These antipatterns arise from distinct traits or behaviors exhibited by specific team members. For instance, a team member who overly assists others might unintentionally foster dependency.	This typically involves tackling the root cause directly, often requiring the person responsible for the antipatterns to take corrective actions.
Practice-related	These antipatterns are recurring and detrimental practices within the engineering process that can negatively impact project outcomes. They often arise from shortcomings in the workflow or from methodologies or communication channels employed by teams.	This often requires course correction, decision making, and communication for the entire team, usually initiated by team leaders.
Structural	These antipatterns revolve around the team's structural elements, such as how team members are organized, roles are assigned, and collaboration is structured. These antipatterns highlight issues such as knowledge silos, communication gaps, and imbalanced skill distribution due to suboptimal team structuring.	By identifying and rectifying structural antipatterns, teams can foster better communication, equitable skill development, and enhanced collaboration, ultimately contributing to a more cohesive and effective working environment.
Leadership	These antipatterns are individual antipatterns that stem from team leaders' behaviors, decisions, or actions. They manifest when leaders exhibit counterproductive traits that hinder team collaboration, impede progress, or create a hostile work environment. For example, a leader who micromanages might inadvertently stifle innovation and demotivate team members.	The challenge with leadership-related antipatterns is that they originate at a pivotal point of influence, affecting team morale and overall project outcomes. Recognizing and rectifying these antipatterns requires a senior team member or senior leadership who can guide the leader to intercede or mediate for the team.

With that background on where the different types of antipatterns can originate and how they can be attacked, let's look at the antipatterns that fall under each category in detail. Some of these antipatterns have been discussed previously in other books or articles with different objectives.[1] I have grouped similar ideas and included my perspective to present antipatterns relevant to team leaders who wish to retain the effectiveness of their teams. Figure 5-1 provides an overview of the different patterns we will discuss under each category.

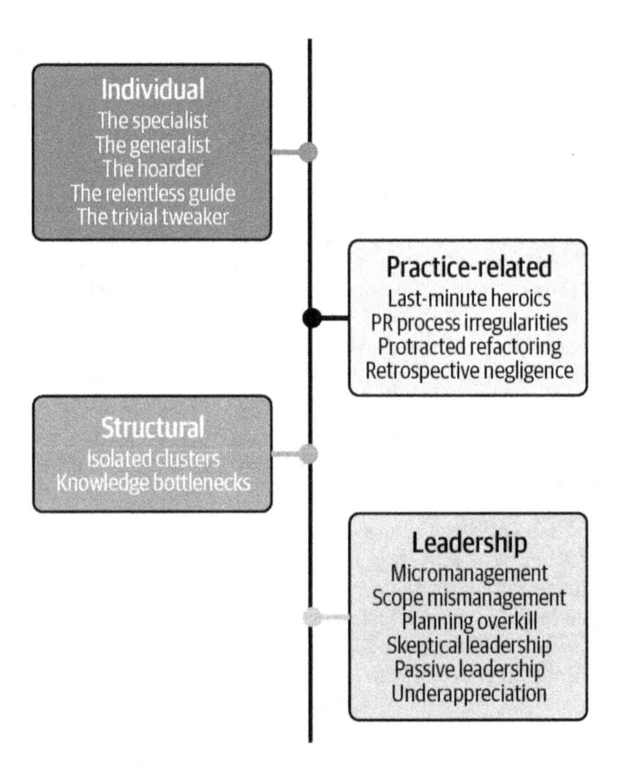

Figure 5-1. Antipatterns overview

1 Other references on software engineering and project management antipatterns: "20 Patterns for Data-Driven Leadership" (*https://oreil.ly/yEH7R*), "Project Management AntiPatterns" (*https://oreil.ly/FtMQG*), and "Five Management Anti-Patterns and Why They Happen" (*https://oreil.ly/NIlfy*).

Individual Antipatterns

Individual antipatterns start and end with an individual. They are born out of personal traits that have their benefits but can also prove detrimental if unchecked. While someone's approach might initially seem efficient and results oriented, it can eventually affect team dynamics adversely. Individuals may often genuinely believe their contributions are beneficial, but unintended consequences can emerge. Addressing these antipatterns typically involves tackling the root cause directly, often requiring the person responsible for them to take corrective actions. The following are the most common antipatterns that fall under this category.

THE SPECIALIST

You have most likely encountered a *specialist* in one of your teams. This person is most strongly identified with a particular module or feature. They have mastered the ins and outs of that part of the codebase, including small hacks that no one knows about. They may or may not have details about the rest of the system, but they certainly know how that specific piece fits. This intricate association might seem advantageous initially, but it can gradually introduce challenges that reverberate across the team and the project.

For example, say your team is developing an online magazine app with a custom widget component that can load different types of content. One of the engineers on the team has not only developed the component but has also handled all the subsequent changes. They know the APIs it uses on the backend, the design for the drag-and-drop features, personalization aspects, etc., right down to the cascading style sheet classes used. They are the only one who knows the special workaround to make it work on an older browser XYZ. This is undoubtedly a *specialist*.

The emergence of specialists is often a consequence of circumstance rather than design. These people were repeatedly assigned to the same module over multiple iterations of the system/product. With time, they became quick in resolving issues in the same area and thus proved to be productive for the team. Such situations can inadvertently transform individuals into go-to experts for that domain. Eventually, they became the module's tacit owners and de facto guardians.

Whether you are new to leading an ongoing project or have been a part of it all along, some of the following telltale signs can help you identify the specialists and their specialization:

- The person has been around long enough to have mastered a specific aspect of the project.
- This is the person the team points you to whenever you ask anything about the particular module.
- Other team members are curious but wary of touching that part of the code and are underequipped to review it.
- The person is confident about the changes they make either to improve the code or to address issues and change requests. The code they submit needs minimal rework.

Having a bunch of specialists on your team may sound like a perfect solution, creating a comfort zone for everyone involved. However, it is a high-risk antipattern that can cause many pitfalls for the team and the specialist:

- Specialists can become indispensable to the team, so much so that the team fears the repercussions of their absence from work for more than a few days. The team believes that the specialist moving to another team or organization would be a nightmare for them all. Thus, the specialist becomes a single point of failure, making the team overly reliant on their presence.
- It's impossible to document the micro-knowledge that the specialist possesses. This can create knowledge silos. Even if knowledge-sharing sessions are conducted, other engineers need hands-on experience with the module before they can begin to understand the business logic or performance optimizations that led to the current state of the code.
- Some specialists may become rigid and overconfident about their knowledge and ignore valuable feedback that can actually improve the usability of their area of expertise.
- Working on the same thing for a long time limits the individual's professional growth. They may never have the time or inclination to acquire new skills, leading to stagnation.

As a team leader or manager, you might get comfortable with the idea of specialization, but it's essential to balance deep expertise with a more holistic and productive team dynamic. While it's good to have experts on the team, it's also necessary that every expert has backups who are at least 60%–80%

as knowledgeable as the expert. Leaders must take specific actions to mitigate the risks due to overspecialization and build a scheme for specialization that's sustainable for their team in the long run:

- Encourage team members to develop expertise in different areas, fostering an adaptive and resilient team.
- Encourage the specialist to delegate some of their changes to other engineers while they take up a reviewer role for these changes.
- Encourage the specialist to document exceptional cases and FAQs related to the module they are experts in. They might also arrange a knowledge-sharing session for these.
- Nudge the specialist to expand their outlook and look for other areas that interest them within the project. They could start with something related to their module so that they are familiar with it.
- During one-on-ones, inquire about the specialist's aspirations, learning goals, and desired training.

The specialist antipattern underscores the importance of cultivating not just experts but a team of adept professionals who can effectively collaborate and support one another. Striking this balance ensures the long-term growth, adaptability, and success of both individuals and the team.

THE GENERALIST

As the name suggests, the *generalist antipattern* arises when an engineer, driven by a desire to diversify their skills and contribute across various domains, spreads themselves too thin and inadvertently compromises depth and expertise in any particular area. While they may intend to become more adaptable, the outcome can be a lack of mastery, focus, and sense of ownership.

Visualize a team working on a diverse project with many modules and layered architecture. Generally, you have structured your team by role—business analysts, backend/frontend engineers, test engineers, etc. However, one engineer stands out as a zealous generalist, eagerly delving into diverse areas and offering assistance wherever needed. The willingness to engage across domains and wear multiple hats might appear virtuous and helpful for the team. Still, it can inadvertently dilute expertise and create an illusion of competence without true mastery.

The generalist's behavior often emerges from a well-intentioned desire to be versatile. Engineers know their ability to contribute across domains can add value to the team. However, for themselves, this pattern can result in a lack of in-depth understanding in any specific area.

Note

Ideally, you want team members to develop a T-shaped skill set (*https://oreil.ly/gQQMu*) where the vertical bar of the T represents their specialization in a specific project area and the horizontal bar represents their ability to collaborate and contribute across different areas of the project.

As a leader, you can easily spot generalists because they are everywhere. Look for the following traits:

- This engineer strives to contribute across various domains, demonstrating versatility.
- The engineer's involvement is marked by breadth, but their depth of understanding is limited.
- Tasks initiated by the engineer are often handed over to specialists for completion.
- The illusion of competence is maintained, but true mastery remains elusive.

By attempting to be versatile, the generalist spreads their cognitive focus across diverse domains, ultimately diminishing their potential to become a true expert in any single area. As a result, tasks requiring specialized knowledge often require the involvement of dedicated specialists, thus subverting the generalist's intended versatility. Moreover, this approach can lead to a sense of superficial competence that masks an absence of in-depth understanding. This illusion can hinder the team's efficiency and pose challenges when dealing with intricate or critical tasks.

To address this antipattern and optimize the team's expertise, team leaders must do the following:

- Task the generalist with a specific area of the project that they can develop mastery in while retaining their general knowledge of the entire project.
- Encourage engineers to focus on honing their expertise in areas that align with their strengths and interests.

- Foster a collaborative environment where generalists collaborate with specialists to achieve optimal outcomes.
- Promote continuous learning and the pursuit of mastery within specialized domains.

In essence, the generalist antipattern illuminates the importance of balance in skill diversification. A thriving team comprises people with a lot of general knowledge about the project for versatility and some specialized ability for depth. By identifying and addressing the tendencies associated with this antipattern, teams can cultivate a well-rounded team dynamic that embraces the strengths of individuals while fostering mastery and expertise.

THE HOARDER

You would usually know what to expect from a specialist or a generalist, but it's different with the next type of individual who can block team effectiveness—the *hoarder*. This individual has a unique modus operandi—quietly working away on various tasks during a sprint, withholding updates until they culminate in one colossal PR. This approach might appear effective if the submitted code integrates perfectly with the rest of the sprint. But, if it does not, you are in for a tedious review session followed by a race to meet the sprint deadline.

Imagine a scenario where your team is engaged in crafting a feature-rich application. One team member seems absorbed in their tasks amid the collective efforts, consistently amassing changes throughout the sprint. Their reasoning may be rooted in a desire to present a comprehensive contribution, or they are probably just fearful of being judged by others for submitting undercooked code. The hoarder may have previously worked on projects that required less collaboration and developed a habit of working in isolation. They may be ignorant of the risks of continuously working as a separate unit.

Whatever their intentions, hoarding code could:

- Indicate that the hoarder does not trust their team
- Signal that the hoarder does not feel safe sharing their work
- Lead to a situation where other team members cannot trust the hoarder

Identifying a hoarder is relatively easy, and you might observe them in the first sprint itself:

- The engineer's updates are withheld until they accumulate into a substantial pull request.
- Other team members find themselves largely unaware of the hoarder's ongoing work.
- Code reviews of the hoarder's sizable pull request become time-consuming and happen so late that you may be unable to devote adequate resources to them, leaving the code under-reviewed.
- The engineer believes their code to be perfect and may not readily accept suggestions to change it to fit better with the rest of the solution.

Hoarding of code could have far-reaching consequences on the deliverables and the team's cohesion:

- The hoarder disrupts the team's collaborative rhythm by consistently working privately. It hampers the continuous feedback loop and impedes the seamless integration of changes. This results in a slower feedback cycle, making it harder to catch and rectify issues early.
- The hoarder becomes a bottleneck during the code review stages. The monumental pull request necessitates more time for a thorough assessment, affecting the overall efficiency of the sprint's closure.
- The isolationist approach also hampers knowledge sharing. Other team members remain uninformed about the ongoing work, missing opportunities to provide insights or collaborate effectively.
- While the hoarder's intentions might be to showcase a substantial contribution, the practice inadvertently stifles the team's agility and cohesion. The antidote lies in fostering a culture of transparency and continuous integration.

As a leader, you can mitigate the hoarder antipattern and avoid running into these issues repeatedly:

- Encourage frequent commits and small pull requests that showcase ongoing progress.

- Promote daily stand-up meetings where team members share their activities.
- Advocate for early and frequent code reviews to ensure continuous integration and minimize bottlenecks.
- Discuss these issues with the hoarder directly but gently. Help them see why their behavior could be detrimental to team efforts.

To give you an example, I had a hoarder on my team once. John was exceptionally skilled, yet he worked in isolation. His first significant project was an important module for Chromium, which he tackled alone. His approach resulted in an intricate piece of code, but it was complex and challenging for others to integrate or understand.

The first lesson I imparted to John was about the power of collaboration. I advised, "Code is read more often than it is written. It should be a narrative for your team, not a solo puzzle." This advice marked the beginning of John's transformation from a lone coder to a collaborative team member. But that was not the end of all problems.

John's next hurdle was effective communication. He often struggled to convey his ideas and missed important cues during team interactions. A notable instance was during a project update, where his overly technical explanations confused most of the team.

I worked with John to hone his communication skills, emphasizing simplicity and clarity. "Explain it as if you're talking to a smart high school student" became our mantra. This not only improved his presentations but also deepened his understanding of his work.

Initially, John was reluctant to share his knowledge, fearing it might lessen his value to the team. This changed when he helped a junior engineer, who was struggling with a bug that John had previously encountered and solved. Encouraging John to mentor the junior colleague was a turning point. He soon realized that sharing knowledge didn't reduce his importance; it actually multiplied it. This realization led him to become a mentor and significantly increased his impact on the team.

A key moment for John was when he led a project that didn't succeed due to a risky technical decision. Unlike before, John accepted responsibility without defensiveness.

Our discussions post-failure focused on learning rather than the mistake. I reminded him, "Every failure is a stepping stone to greater success." This mindset shift was critical, as it taught John to embrace risks and learn from setbacks, a vital quality for any technical leader.

Today, John is more than a technical leader: he's a mentor and an inspiration to his colleagues. He exemplifies the values of collaboration, clear communication, and continual learning. The ultimate lesson for John was understanding that leadership in software engineering is about empowering others. It's about creating a team that's resilient, adaptive, and innovative.

The hoarder antipattern teaches us the importance of consistent collaboration and transparency. A successful team thrives on collective efforts, continuous communication, and the incremental integration of changes. By recognizing and addressing hoarding tendencies, teams can ensure smoother workflows, faster code reviews, and a healthier team dynamic.

THE RELENTLESS GUIDE

Some software engineering teams encounter the *relentless guide antipattern*, characterized by an engineer's eagerness to offer assistance beyond its intended scope. They support others with good intentions and help to improve the quality of code. You might find this unofficial mentor admirable initially. However, it becomes a problem if one or more engineers demonstrate excessive reliance on their relentless guide and cannot make decisions on their own.

Imagine a team united in its pursuit of a complex project. Amid this collaboration, an engineer stands out as a relentless guide, consistently providing solutions, guidance, and answers whenever team members encounter challenges. Code review sessions often become extensive redesign or rewrite sessions driven by the guide. Every time one of the mentees tries to solve a problem by themselves, they are spoon-fed the perfect solution before they have a chance to analyze the situation. While well-intentioned, this dedication can subtly cultivate a culture where individuals frequently turn to the helper for answers rather than developing their problem-solving capabilities.

The relentless guide antipattern typically emerges without ill intentions. Engineers often engage in this behavior to cultivate a supportive environment. As a community, we encourage senior engineers to help junior engineers apply their skills in real-world scenarios. Yet, at its extreme, the result can hinder the natural progression of skills within the team, leading to a lack of self-sufficiency.

Identifying this pattern involves observing specific traits of the relentless guide and their team:

- The engineer eagerly offers assistance without being prompted.
- Team members habitually seek the engineer's guidance, even for minor matters.
- Newcomers to the team rapidly become reliant on the engineer's expertise.
- The engineer is confident and effective in addressing challenges, often preemptively.

However, beneath the surface lies a series of pitfalls that affect not only the team's overall growth but also the helper's effectiveness:

- By assuming the role of an omnipresent problem solver, the guide inadvertently hampers the team's individual and collective development. Team members become accustomed to relying on external support rather than sharpening their skills and knowledge.
- Furthermore, the helper's time and energy become disproportionately consumed by continuous assistance, leaving less room for personal productivity and responsibilities. This imbalance can lead to burnout and hinder the team's overall progress.
- While the helper's enthusiasm is commendable, the outcome hampers the team's autonomous learning and self-sufficiency potential. The solution lies in cultivating a balanced learning environment that encourages helpfulness and independent exploration.
- If a few team members have become dependent on a guide, it might split the team into two factions: one that agrees with the guide by default and another that questions their reasoning.

To mitigate the relentless guide antipattern and promote well-rounded growth, do the following things:

- Encourage team members to attempt problem-solving independently before seeking aid.
- Foster peer learning by pairing individuals with different experience levels.
- Organize regular knowledge-sharing sessions to encourage diverse learning.

- Urge the helper to allocate dedicated time for their own tasks and development.
- Assign the guide to challenging tasks where they may be required to devote time to trying something unfamiliar.

The relentless guide antipattern reminds us to balance helpfulness and self-driven learning. A thriving team embraces shared expertise, embraces challenges collectively, and ensures each member's progress is nurtured. By preventing themselves from falling prey to the relentless guide antipattern, teams can harness the true potential of both helpers and those seeking assistance.

THE TRIVIAL TWEAKER

The *trivial tweaker antipattern* is one of the more subtle individual antipatterns. These engineers (and their code) do not stand out like a relentless guide or a specialist, nor do they hide like the hoarder. Instead, they regularly deliver small but insignificant changes.

This pattern emerges when engineers, driven by moments of monotony or a desire for perceived code improvement, engage in small and insignificant code changes under the pretext of refactoring or reorganizing. Although these adjustments might seem innocuous, they can lead to unproductive diversions, misallocation of resources, and a dilution of focus from meaningful tasks. The changes often lack significant impact and might create a culture of undue attention to trivial adjustments.

Identifying this antipattern involves recognizing its characteristics:

- An engineer consistently indulges in minor code alterations that produce little substantial value for the stakeholders.
- You find perfectly stable and optimized code getting refactored with little change in performance or usability for the product.
- The team's discussions may be sidetracked by deliberations over changes, and team members realize a little late that they make little difference to the project outcomes.
- Valuable time is spent reviewing the trivial tweaker's superficial adjustments.
- The line between meaningful refactoring and inconsequential tweaking is blurred.

While not malevolent, this pattern introduces inefficiencies and missed opportunities for the individual's growth and achievement. By allowing themselves to be entangled in inconsequential refactoring, the trivial tweaker diverts cognitive energy from tasks that could meaningfully advance the project. They are either unclear about project goals or are not motivated to care about them.

Unnecessary refactoring of stable code may also lead to issues if the person changing the code is not fully aware of all the business logic in the code and its repercussions.

As a team leader, you must help the trivial tweaker get back on track. Some ways of doing this are as follows:

- Assign a new and challenging task to the trivial tweaker so they have to create something from scratch rather than tweak the existing code.
- Encourage the engineer to critically evaluate the potential impact of code changes before diving into refactoring.
- Foster an environment that values substantial contributions and meaningful improvements over superficial tweaks.
- Establish clear criteria for code alterations to align with the project's objectives.

An effective team focuses on tasks that yield tangible progress, avoiding distractions arising from an undue fixation on minor changes. By identifying and mitigating the tendencies associated with the trivial tweaker antipattern, teams can enhance efficiency, maintain focus, and achieve their goals more effectively.

Practice-Related Antipatterns

Practice-related antipatterns are caused by procedural pitfalls when designing, coding, reviewing, or testing solutions. A single individual cannot be held responsible for these. They can only be managed when the entire team collaborates to standardize and follow processes under the guidance of a leader.

LAST-MINUTE HEROICS

Last-minute heroics is a practice-related antipattern that can become a trend within a software development team. Issues and challenges are often addressed hastily and heroically just before a release, with minimal time for comprehensive feedback or testing. While such heroic saves might seem impressive and efficient, they can mask underlying problems and introduce unnecessary risks to the project.

Imagine you have a team engrossed in developing a software product, aiming for a release. As the deadline approaches, some of your junior developers struggle to address bugs identified during code reviews and testing. Your team's generalists and relentless guide figures may jump in to help them. In a display of remarkable effort, team members rally to fix these problems at the eleventh hour to prevent potential delays. Eventually, the test team retests the issues and approves the fixes applied. The efforts are celebrated, and everyone is happy. But then, the cycle repeats!

The allure of last-minute heroics often emerges from a desire to meet deadlines and ensure a successful release. But imagine if this were to become a pattern repeated before every release or at the end of every sprint. Relying on heroic interventions every time signals that there is something wrong at the process level, and it can be risky for the project and lead to several challenges:

Lack of feedback
: Rushing to address issues before release leaves little time for thorough testing and feedback, leading to potential oversights or incomplete solutions.

Hidden technical debt
: Quick fixes might patch immediate problems but could accumulate technical debt, creating more complex issues in the long run.

Decreased quality
: These patches can undermine the overall quality of the product, eroding user trust and satisfaction.

Dependency on heroes
: Team members might become dependent on the notion that last-minute saves will always rescue them, hindering the establishment of robust processes.

Mitigating the last-minute heroics antipattern requires a shift toward proactive, well-defined processes:

Effective planning
: Ensure that you have a thorough plan, regular checkpoints, and continuous testing throughout the development cycle to reduce the likelihood of significant issues arising late in the process.

Transparent communication
: Encourage open communication about challenges, enabling early identification and resolution of potential roadblocks.

Prioritized backlog
: Maintain a prioritized backlog that systematically addresses critical issues and features, preventing the accumulation of unresolved problems.

Sustainable pace
: Avoid pushing teams to their limits in a race to the finish line, as it can lead to burnout and subpar outcomes.

In conclusion, last-minute heroics might provide temporary relief, but it is an unsustainable approach that hampers long-term success. Embracing proactive, well-defined processes ensures a higher-quality product, minimizes technical debt, and cultivates a culture of consistent excellence.

PR PROCESS IRREGULARITIES

The *PR process irregularities antipattern* highlights various challenges within a software development team's pull request (*https://oreil.ly/SEV5S*) (PR) or code review workflow. Let's talk about a few distinct irregularities in the PR workflow, which can introduce inefficiencies, decrease code quality, and hinder collaboration:

Rubber-stamping
: PRs are approved without adequate review in this scenario. Quick approvals might seem efficient, but they can overlook critical issues, leading to potential bugs and suboptimal code quality. Rubber-stamping diminishes the value of the review process and weakens the collaborative essence of PRs.

Self-merging
: When engineers approve their own PRs, it creates a conflict of interest and bypasses the crucial checks and balances that reviews provide. Self-merging can result in a lack of diverse perspectives, reducing the chances of catching errors or suggesting improvements.

Long-running PRs
: Long-running PRs, constantly going back and forth between the engineer and the reviewer, signify inefficiencies in communication and decision

making. Delays in addressing issues can prolong the integration process, slowing the entire development cycle and impeding progress.

Last-minute PRs

Last-minute PRs submitted by multiple engineers just before a deadline signal poor planning or estimation by engineers. Such PRs can pose risks if rushed and not given the same due diligence as earlier PRs. Such trends can also mean there were bottlenecks in the execution, because of which the work was delayed.

Addressing these problems requires establishing a robust PR process that promotes thorough review, accountability, and timely progression:

Thorough review and accountability

Mandate a review process that necessitates a comprehensive evaluation of each PR. Encourage open discussions, allowing reviewers to provide constructive feedback and suggest improvements. This practice reduces the risk of rubber-stamping and encourages active participation in the review process.

Diverse approvals

Implement a rule that prevents self-merging. Instead, enforce a practice where PRs require approval from different team members, ensuring a wider range of perspectives and increasing the chances of catching errors before they reach the codebase.

Timely feedback and closure

Set clear expectations for timely PR reviews and responses. Introduce mechanisms to avoid PRs becoming long running. Foster communication between engineers and reviewers to promptly address concerns and reach a resolution. This approach accelerates decision making and prevents unnecessary delays.

Intermediate checkpoints

Ensure you have established checkpoints to review progress and identify bottlenecks. This could be daily stand-ups where engineers can update each other on progress or scheduled meetings to discuss issues or change requests that are delaying their progress.

Code Reviews at Google

A 2018 case study on modern code review at Google (*https://oreil.ly/JqAj9*) revealed the following best practices that have been iteratively refined through millions of code reviews carried out over more than a decade of code changes:

- Education, maintaining norms, gatekeeping, and accident prevention are the key expectations from the code review process.
- Code reviews enable future auditing through the tracking history generated as part of the process.
- Expectation for a specific review depends on the work relationship between author and reviewer. For example, education is the primary objective when the project lead reviews the code, but during peer reviews, the focus tends to be on accident prevention.
- The median developer authors about three changes a week, and the median time for initial feedback is under an hour for small changes and about five hours for very large changes.
- The code review process is lightweight and flexible with smaller changes and quicker reviews. The median reviewer count per change is one. The number of large/complex changes requiring more than one reviewer is less than 25%.
- The number of review comments is higher for people new to Google but stabilizes at around four comments per hundred lines of code for older employees.
- Breakdown in the process is largely due to the interactions during reviews: for example, misunderstandings about why the change was required.

PR process irregularities can hinder a team's efficiency, code quality, and collaboration. A well-defined PR process counteracts these issues by fostering comprehensive reviews, diverse approvals, and timely resolutions. By establishing a coherent PR process that addresses these irregularities, teams can reap numerous benefits:

- Code quality improves as issues are identified and resolved early in the process.
- Collaboration is enhanced through diverse reviews and valuable feedback from team members.
- Efficiency increases as timely PR reviews prevent bottlenecks and promote consistent progress.

Embracing this approach nurtures a culture of accountability, quality, and efficient development, leading to a more successful and cohesive software development lifecycle.

PROTRACTED REFACTORING

While working on a project, self-motivated engineers often discover existing code that could be written better, for example, by using a new API or library that was not available previously. They might refactor/rewrite the code themselves on the go. With approvals, a refactor of substantial size could also become a mini-project involving a few team members.

A *protracted refactoring antipattern* is characterized by a code refactor that transforms into an enduring process involving one or many engineers. The refactor stretches beyond its expected timeline, either driven by perfectionism or insufficient domain knowledge, leading to a long-running cycle of adjustments beyond initial intentions. As the refactor expands in scope and complexity, it can impede project progress, hamper focus, and foster inefficiencies.

Engineers may initiate a deliberate code refactor to elevate software quality. A few engineers collaborate, each contributing insights and improvements. However, the planned refactoring evolves into a protracted cycle, stretching over time and drawing in additional engineers. What started as a genuine desire to enhance code quality and maintainability via meticulous refining may eventually become a complicated project in itself if it is not reined in at the right time. The existence of the antipattern is characterized by the following:

Escalating scope

The planned refactor expands organically, incorporating more elements and aspects, leading to unintended scope expansion.

Progress delays

As the refactor persists, project timelines may extend, potentially delaying releases and affecting overall project objectives.

Resource drain
: Multiple engineers enmeshed in the prolonged refactoring divert valuable resources from other essential tasks.

Diluted focus
: Constant refactoring siphons energy and time away from critical tasks, undermining the team's overall effectiveness.

To mitigate the protracted refactoring antipattern, a balanced strategy is vital. Consider doing the following:

Identify the cause
: Identify what is causing the protraction:

  - It could be a desire for perfection, in which case you must help the engineer or team clearly define the scope of refactoring. Establish precise boundaries for the planned refactor, outlining objectives and areas of focus to avoid scope creep.
  - Another issue could be insufficient domain knowledge. The engineers may not be sure of the purpose of certain blocks of code, and it could be risky to refactor them without this knowledge. In this case, support them by discussing what you know or connect them to an engineer who has previously worked on that part of the code.

Set time constraints
: After identifying the cause and a suitable solution, allocate realistic timeframes for the refactor, emphasizing efficiency and maintaining project momentum.

Peer review and closure
: Implement regular peer reviews to gauge the refactor's impact and decide when the process should conclude.

Open communication
: Foster open dialogue to ensure the refactor remains aligned with project goals and does not extend indefinitely.

In conclusion, refactoring may stem from a well-intentioned pursuit of excellence and collaboration, but it can lead to unintended complications if it stretches beyond the initial scope. By defining clear scope, adhering to time constraints, and maintaining effective communication, teams can harness the benefits of refactoring while upholding project objectives. Addressing the complexities associated with this antipattern empowers engineers to balance meaningful improvements and the project's overall advancement.

RETROSPECTIVE NEGLIGENCE

In agile projects, retrospective meetings that are held at the end of an iteration provide teams with a dedicated space to reflect on their work processes, communication, and collaboration. These structured sessions offer an opportunity to identify what went well, what could be improved, and how to enhance future performance. Retrospectives are pivotal in fostering continuous improvement, enhancing team dynamics, and driving innovation within a software development project.

However, retrospectives can become ineffective if the proper process is not followed while conducting them. The *retrospective negligence antipattern* reveals irregularities in the conduct and effectiveness of retrospectives, impeding their potential benefits. Several issues can contribute to the diminished effectiveness of retrospectives:

Missed/delayed retrospectives
: When teams skip retrospectives due to time constraints or other priorities, they miss the opportunity to reflect and improve upon their processes and collaboration. Similarly, if they delay a retrospective, they may miss important details that are fresh in everyone's mind just after a sprint or an iteration is completed.

Shortened sessions
: Holding brief retrospectives that lack sufficient time for meaningful discussion can prevent the team from delving deep into their challenges.

Lack of structure
: Without a defined framework, retrospectives might become unstructured discussions that fail to identify critical issues and potential solutions.

Unanimous agreement
: If team members avoid discussing conflicting viewpoints or only a few dominant voices lead the conversation, actual areas for improvement may remain hidden.

Lack of follow-up
: Not following up on the action items or improvements identified during retrospectives can result in a lack of accountability and render them meaningless.

Surface-level analysis
: Addressing only surface-level symptoms rather than exploring the root causes of challenges can limit the effectiveness of retrospectives.

To counter the retrospective negligence antipattern and maximize the value of retrospectives, teams must understand why retrospectives are essential and what they can gain from them. Teams can ensure that irregularities don't creep into the retrospective process by doing the following:

Prioritize regularity
: Make retrospectives a consistent part of your project's workflow, allocating time for them at the end of each iteration or milestone.

Allocate adequate time
: Ensure retrospectives have ample time for a thorough discussion, allowing team members to explore issues and solutions in depth.

Embrace structure
: Implement well-defined retrospective frameworks such as Start, Stop, Continue (*https://oreil.ly/aIz4N*) or Mad Sad Glad (*https://oreil.ly/sEoal*) to guide discussions and ensure comprehensive analysis.

Encourage diverse participation
: Create a psychologically safe environment where all team members feel comfortable sharing their perspectives and facilitators ensure balanced participation.

Implement action items
: Ensure that the identified action items are documented, assigned to responsible individuals, and followed up on in subsequent retrospectives.

Focus on root causes
: Encourage the team to probe beyond surface-level issues and uncover underlying causes to address challenges at their source. Use techniques like a five whys exercise (*https://oreil.ly/_A5B3*) or a fishbone analysis (*https://oreil.ly/QO1Go*) to get to the root of the problem.

By adhering to these best practices and nurturing a culture of open communication and continuous improvement, teams can counter the retrospective negligence antipattern. These regular reflections empower teams to refine their processes, enhance collaboration, and evolve their practices to achieve higher levels of performance and success.

Structural Antipatterns

Structural antipatterns are caused by structural issues, including knowledge silos, skill imbalances, and communication gaps, which can pose risks to the long-term sustainability of the team.

ISOLATED CLUSTERS

The *isolated clusters antipattern* delves into a situation where subteams or groups form within a larger software development team, leading to insular pockets of collaboration. Within these clusters, members predominantly engage with colleagues from their own subgroups for reviews, help, or informal knowledge sharing. This inadvertently creates barriers to cross-cluster collaboration and hinders the team's collective progress.

As a project evolves, smaller clusters naturally emerge, with members gravitating toward those with shared interests or responsibilities. These clusters engage in focused discussions and collaboration, which can inadvertently lead to a lack of interaction with members from other areas. They may always end up reviewing each other's PRs, leading to a total absence of reviews from engineers outside the cluster.

There is comfort and efficiency in working closely with familiar colleagues with whom engineers have established chemistry. While this pattern can lead to enhanced productivity within the subgroups, it can also lead to a few challenges if you do not shake things up occasionally:

Knowledge fragmentation
: As clusters remain focused on specific areas, knowledge sharing across domains becomes limited, leading to gaps in understanding and expertise.

Missed insights
: Friends working closely together might end up rubber-stamping each other's work instead of doing an in-depth review. The absence of interactions with members from other subgroups results in a lack of diverse perspectives, potentially leading to tunnel vision in problem-solving.

Stagnant growth
Isolated collaboration restricts professional growth opportunities, preventing engineers from broadening their skill sets by engaging with different domains.

Reduced cohesion
As subteams grow insular, the sense of a united team can weaken, potentially affecting morale and overall camaraderie.

To mitigate the isolated clusters antipattern and promote holistic collaboration, team leaders must encourage the following measures:

Interdisciplinary sessions
Organize regular sessions that bring members from different subgroups together to discuss projects, encouraging cross-domain interactions.

Rotating roles
Introduce rotation within clusters, allowing members to temporarily engage with different subteams, fostering a well-rounded perspective.

Cross-domain initiatives
Develop projects or tasks that require collaboration between members from different domains, fostering teamwork and shared learning.

Open communication channels
Promote open channels for sharing updates, insights, and challenges across the entire team, ensuring a culture of transparency and knowledge exchange.

The isolated clusters antipattern can unintentionally segregate team members and hinder cross-domain collaboration. By fostering an environment of interdisciplinary engagement, rotation, and shared initiatives, teams can harness the collective potential of all members. Addressing the tendencies associated with this antipattern cultivates a sense of unity and collaboration.

KNOWLEDGE BOTTLENECKS

The *knowledge bottlenecks antipattern* highlights a scenario where vital knowledge and expertise become concentrated within a limited number of individuals, resulting in a low bus factor (*https://oreil.ly/VdXuo*). The *bus factor* is the minimum number of team members required to be incapacitated (hit by a metaphorical bus) before a project faces severe disruption. The pattern often emerges when certain individuals, like the specialist or the relentless guide, become sole repositories of crucial information, creating vulnerabilities for the project's continuity and stability.

Knowledge bottlenecks are represented by a few individuals who are regarded as the ultimate authorities within specific domains. These experts (such as the specialist with intricate knowledge about a module and the relentless guide who consistently guides the team) are considered indispensable resources. However, this concentration of expertise can lead to several challenges:

Single points of failure
: If these individuals were to leave the team or become unavailable, their absence could disrupt the project and hinder its progress.

Dependency
: Team members may become overly reliant on the expertise of these individuals, stunting their own growth and problem-solving abilities.

Knowledge silos
: Information becomes trapped within a limited group, preventing the broader team from understanding and contributing to critical areas.

Communication gaps
: Lack of knowledge distribution can lead to miscommunication and misunderstandings between different project parts.

I had to deal with this antipattern myself in my early days as a leader at Chrome. As our codebase grew in complexity, I noticed pockets of specialized knowledge developing—a few engineers would become experts on particular components and areas of the project, but others had little insight into how they worked. This created some bottlenecks, especially if there were ever interrelated dependencies or the engineer with the specialized knowledge was out of office.

To address this, we encouraged more documentation, sharing of code ownership, and internal tech talks. We also instituted cross-training and rotation programs. Recently onboarded junior engineers were paired with veterans to learn from their deep expertise.

This knowledge sharing broke down silos and opened the bottlenecks. Expertise got commoditized and institutionalized within the team rather than remaining concentrated with specific people. Our team's capabilities scaled exponentially, despite no headcount growth.

Based on this experience, I can advise that to mitigate the knowledge bottlenecks antipattern and ensure a healthy knowledge distribution, team leaders should consistently promote the following activities:

Cross-training
: Encourage team members to collaborate and cross-train across different domains to broaden their understanding and skill set.

Knowledge sharing
: Establish a culture of regular knowledge-sharing sessions or documentation to disseminate critical information among team members.

Pair programming
: Encourage pairing between experienced and less experienced team members to foster mutual learning and knowledge exchange.

Rotation of responsibilities
: Rotate responsibilities periodically to ensure that knowledge is spread across the team and not owned by just a few individuals.

Mentorship
: Introduce mentorship programs where experienced team members guide newer members, enhancing skill development and knowledge distribution.

As you can see, the knowledge bottlenecks antipattern stems from the intention to optimize expertise, but it poses risks to project continuity and team development. By fostering knowledge distribution through cross-training, sharing sessions, and mentorship, teams can break down silos, enhance collaboration, and fortify their ability to weather personnel changes. You thus ensure a high bus factor for the team, implying that a large number of team members would need to be hit by the hypothetical bus before a project faces severe disruption. Addressing the tendencies associated with this antipattern ensures that the project's knowledge landscape remains dynamic, diverse, and resilient.

Leadership Antipatterns

Your actions as an engineering leader can influence most patterns and antipatterns in software engineering teams. But there are cases where the leader's actions themselves become an antipattern. Leadership antipatterns are caused by leadership behaviors and strategies that can hinder team dynamics and project success. These antipatterns highlight issues such as micromanagement, lack of clear vision, and resistance to feedback, which can lead to demotivated teams and suboptimal outcomes.

MICROMANAGEMENT

Micromanagement is a much-talked-about leadership antipattern characterized by excessive supervision, where managers exert unnecessary control and oversight over their teams. Although you may be familiar with the concept, you need to understand the various ways it can manifest itself so that you can consciously stop yourself from falling into this particular leadership trap:

Perfectionist bottlenecks
: Some managers, driven by an obsession for perfection, continually introduce new elements or changes from an extensive wish list. While this may enhance quality, it often slows progress, causing delays and creating unnecessary complexity.

Prescriptive direction
: Certain managers tell their teams exactly how to execute tasks rather than highlighting what needs to be done. This approach stifles the team's autonomy and creativity and hampers their ability to find innovative solutions.

Guardians of information
: Managers can become information gatekeepers, withholding critical details from their teams to shield them from upper management or stakeholders. This withholding of information hinders transparency, impedes informed decision making, and obstructs the team's ability to align with broader organizational goals.

Leaders may not realize that they are doing it, but the harms of micromanagement are significant and far-reaching:

Stifled innovation
: Excessive control curbs the team's ability to explore innovative ideas and alternative approaches. It hampers experimentation and restrains the team from finding novel solutions to challenges.

Low morale
: Overly controlling management erodes team morale. When team members feel their contributions are consistently doubted or disregarded, their motivation and engagement decline, impacting overall job satisfaction.

Slowed progress
: Micromanagement introduces unnecessary bottlenecks. The constant need for approval and detailed oversight elongates decision-making processes, leading to missed deadlines and reduced productivity.

Limited growth
: Autonomy fosters skill development and learning. Micromanagement denies team members the chance to make independent decisions and learn from their experiences, which restricts their professional growth.

Uninformed decisions
: While managers who shield their teams from external drama eliminate distractions, some transparency is required for teams to make the right decisions that align with organizational initiatives.

If you are one of the leaders experiencing any of these symptoms, granting some degree of autonomy to your team can provide relief in the following ways:

Increased ownership
: Autonomy breeds a sense of ownership and responsibility. When trusted to make decisions, teams feel a more profound commitment to their work and take pride in the results.

Enhanced creativity
: An autonomous environment encourages creativity and exploration. Teams are likely to experiment with new ideas and approaches without rigid control.

Improved morale
: A leadership style that respects autonomy fosters positive morale. When team members feel valued and empowered, it contributes to a supportive work atmosphere and heightened job satisfaction.

Faster innovation
: Empowering teams with decision-making authority streamlines the decision-making process. This agility enables teams to adapt quickly, fostering a more innovative and flexible environment.

Glass barriers
: Leaders can act as glass barriers, balancing transparency while shielding their teams. They can provide enough clarity to keep their teams informed about essential external factors that affect their work while effectively shielding them from unnecessary interactions that can waste time or distract them from focused work.

Embracing a leadership approach that grants teams the autonomy to make informed decisions while providing guidance and support is pivotal. This balance facilitates an environment where teams thrive, driving innovation, bolstering morale, and achieving project success.

SCOPE MISMANAGEMENT

Scope mismanagement is a leadership antipattern in which engineering leaders struggle to manage and control the scope of a project. This predicament often arises due to the influx of continuous change requests from product owners or stakeholders, leading to an ever-expanding workload and backlog. As a result, the project becomes mired in uncertainty, hampering progress and causing undue stress for the team, hindering its ability to deliver effectively.

A case study (*https://oreil.ly/6GE8h*) published by the Project Management Institute (*https://www.pmi.org*) talks about how a new document imaging system that was to be delivered in 18–24 months took four years to develop due to scope mismanagement. One of the key reasons identified was that there were about 50 users providing requirements to the vendor's development team and they couldn't agree on their needs. The managers on both the stakeholders' side and the vendors' side were unable to provide any guidance or leadership on managing the scope of the project. Sound leadership on both sides could have helped this project get back on track sooner.

An engineering project should have a well-defined scope and roadmap. While a few change requests are common and reflect well on the team, engineering leaders should know when to stop entertaining such requests. When frequent change requests from various sources accumulate, gradually altering the project's focus and objectives, the workload or the product backlog can spiral out of control. Some of the challenges encountered are as follows:

Incessant change requests
: Frequent and often unstructured change requests from product owners or stakeholders disrupt the project's established direction and priorities, causing confusion and a lack of focus.

Lack of prioritization
: Without a precise mechanism for prioritizing and managing new requests, the project's scope becomes inflated, making it challenging to allocate resources efficiently.

Inflated workload
: The accumulation of additional tasks and features augments the team's workload, leading to burnout, missed deadlines, and reduced productivity.

Delayed deliverables
: The absence of effective scope management results in delayed deliverables and an inability to meet project milestones, eroding stakeholder trust.

Reduced quality
: As the scope widens unchecked, the team may sacrifice quality to meet increasing demands, compromising the final product's integrity.

If you are grappling with inflating scope of your project, there are a few steps you can take to regain control over the project scope:

Change evaluation process
: Implement a structured process for evaluating against a baseline (usually the initial project requirements) and prioritizing change requests, involving stakeholders and subject matter experts to ensure alignment with project goals.

Effective communication
: Foster open and transparent communication with product owners and stakeholders, ensuring they understand the implications of each change request on scope, timeline, and resources.

Scope freeze periods
: Introduce controlled periods during which scope changes are limited to allow the team to focus on completing existing tasks and stabilizing project progress.

Regular reviews
: Conduct regular scope reviews to assess the impact of changes and make informed decisions about incorporating new features or adjustments.

Empowered decision making
: Equip senior engineers with the authority to evaluate the feasibility and impact of change requests, ensuring they align with project objectives.

Escalation
: If open communication with product owners does not help and the scope remains volatile, escalate to senior management.

Scope mismanagement can undermine project progress, team morale, and overall success. By implementing robust processes, effective communication, and proactive scope management strategies, engineering leaders can navigate the challenges of scope creep, maintain project focus, and deliver results that meet both stakeholder expectations and project goals.

PLANNING OVERKILL

The *planning overkill antipattern* characterizes a leadership approach often seen in waterfall software development (*https://oreil.ly/ioBfL*), where excessive time and effort are invested in the planning and analysis phases. While comprehensive planning and design are necessary, this pattern veers to an extreme, attempting to account for every conceivable detail up front. Consequently, the team becomes paralyzed in a cycle of seeking perfection, which stifles progress and hampers the ability to move forward effectively.

Imagine a team working on an ecommerce application development project. The stakeholder has tasked the team with implementing a feature that calculates discounts for various products during checkout. Accuracy is critical because incorrect discounts could result in customer dissatisfaction and financial losses. As the team members dive into the feature, they start overanalyzing every aspect. The team spends weeks analyzing different discount scenarios, considering various pricing models. The team comes up with a highly configurable design so that administrators could apply different strategies to products over specific time periods, making the design very complex. The leader becomes deeply involved in the implementation, offering numerous suggestions for improvement. The stakeholder probably expected something much more straightforward, such as standard discounts for all products.

In the planning overkill antipattern, the team's progress is hampered by several challenges:

Overanalysis
: The team dedicates an inordinate amount of time to dissecting and analyzing requirements to an exhaustive degree, with no user interaction and much speculation.

Endless design iterations
: Striving for perfection leads to multiple design iterations, each attempting to anticipate every possible scenario before execution begins.

Extensive documentation
: If the analysis-and-design phase has been extensive, the team must create comprehensive documentation for its plans, assumptions, constraints, and design decisions for future reference. The more you plan, the more you must document, further delaying the development.

Delayed development
: Excessive time spent on planning and design results in delayed development start, causing setbacks and extending the project timeline. Speculating about extreme conditions during the design phase leads to a system that is difficult to develop, document, and test.

Inflexibility
: The rigid adherence to predefined plans makes it challenging to adapt to changing requirements or unforeseen challenges during development.

While project planning is essential for effective management, to promote a balanced approach, you can use the following techniques:

Realistic scope
: Define a scope that allows for necessary planning and design while acknowledging that unforeseen factors will emerge during development.

Iterative refinement
: Adopt an iterative approach to planning and design, allowing for adjustments based on new insights and feedback. Kent Beck (*https://oreil.ly/zhtv7*), creator of the Extreme Programming methodology, is known to have said, "An iteration's worth of data is worth months of speculation."

Flexibility in execution
: Embrace the fact that not everything can be predicted up front and allow the team to adapt as the project progresses.

Progressive elaboration
: Begin with foundational planning and gradually refine details as development unfolds, avoiding the need for exhaustive up-front analysis.

Risk management
: Focus on addressing high-priority risks and challenges while understanding that some risks can only be genuinely understood during execution.

Although the planning overkill antipattern originates from the intent to achieve flawless outcomes, it can hinder progress and responsiveness. Balancing planning with the acknowledgment of the unknown is vital. Embracing a mindset that allows for adaptive planning, iterative refinement, and a willingness to navigate uncertainties leads to a more agile and effective software development process.

SKEPTICAL LEADERSHIP

Skeptical leadership signifies an antipattern where engineering managers develop unwarranted insecurities about the team's competence as a project progresses. These anxieties could appear as irrational fears about things that could stop working. Every minor issue looks like a potential minefield to a skeptical leader. A random media report or anecdote from another team might lead to unnecessary questions about design or technology choices.

Managers have the right to raise these questions and be concerned about the project's direction. In most cases, the manager's fears originate from a place of commitment to the project and concern for its health. Healthy skepticism based on the principle of "trust, but verify" (*https://oreil.ly/NQYGd*) may indeed be beneficial for a project. But when these fears are voiced offensively or suspiciously, they can undermine the team's autonomy and confidence.

Imagine a team where the manager constantly questions various aspects of the project. For example, during a sprint review, when a developer explains that a particular feature took longer to implement due to unforeseen technical challenges, the manager responds, "Why did it take so long? We discussed this weeks ago." Another time, when the team proposes using React as a suitable library for their frontend project, the manager asks, "Why are you using React? Isn't Vue.js more lightweight and easier to learn?" In these scenarios, developers must dedicate excessive time to addressing managerial concerns, diverting focus from core technical responsibilities.

The insecurities of a skeptical leader can manifest themselves in several ways, leading to challenges for the team:

Unfounded fears
: Managers develop fears about possible issues late in the project, despite the team's competence and track record.

Insecure technology decisions
: These irrational fears may result in managers advocating for technological changes or approaches based on misinformation or hearsay.

Passing the pain
: Leaders may experience skepticism about their teams from senior management and may pass the questions to their engineering teams.

Constant reassurances
: Developers spend considerable time responding to and addressing managerial concerns, detracting from their primary technical tasks.

Diminished confidence
: As managers voice doubts, team confidence may waver, impacting morale and overall productivity.

Slowed progress
: The need to provide constant explanations and reassurances slows development, leading to missed milestones and extended timelines.

As a leader, you must identify these tendencies in yourself and other leaders and act accordingly to restore productive collaboration:

Evidence-based decisions
: Before approaching the team, ensure your concerns are based on factual information and evidence.

Effective communication
: If you think your concerns are valid, liaise with a single senior developer who can provide explanations and make decisions if necessary. Listen attentively to what they have to say.

Transparency

If a senior leader is likely to question you, ensure that technical decisions are transparently communicated to you and that the rationale behind choices is shared so that you can defend those choices when challenged.

Time management

Allocate dedicated periods for addressing concerns, enabling developers to focus on their core responsibilities during the majority of their work time.

Confidence building

Promote initiatives that bolster the team's confidence, such as showcasing successful past projects and highlighting the team's strengths.

The skeptical leadership antipattern has the potential to erode confidence, disrupt development, and hinder overall progress. By promoting evidence-based decision making, transparent communication, and building trust through open dialogue, teams can overcome unwarranted concerns and maintain a productive and harmonious work environment.

PASSIVE LEADERSHIP

Passive leadership is an antipattern characterized by leaders who exhibit complacency, timidity, and indecisiveness. These leaders are hesitant to drive essential improvements or changes, offering insufficient guidance and actionable feedback to their teams. They favor maintaining the status quo, avoiding disruptions, and sidestepping critical decisions that could drive progress and innovation. They assume that they are likable to their teams when, in reality, their team members like being challenged occasionally.

Passive leaders are also known as absentee leaders. An article in the *Harvard Business Review* identifies these to be the most common type of incompetent leaders (*https://oreil.ly/tsuuC*). The article highlights how absentee leaders often go unnoticed in organizations due to their subtle negative impact. Unlike overtly destructive managers, absentee leaders are not easily detected, allowing them to hinder organizational progress, block succession paths, and contribute little to productivity over time, largely unchecked.

Passive leadership is not sustainable for the following reasons:

Stagnation
: Leaders' reluctance to advocate for necessary improvements or changes stagnates the team's growth and prevents them from adapting to evolving circumstances.

Lack of direction
: The lack of clear direction and actionable feedback confuses team members, affecting morale and diminishing productivity. This refers to managers who always say, "You're doing a great job!" when you are well aware that you are not.

Missed opportunities
: Opportunities for innovation and efficiency enhancements are often overlooked as leaders are reluctant to explore new paths. Team members may come up with plans to improve the existing code or suggest a new automation that would improve developer experience. However, the manager is averse to trying something new due to a fear of breaking something that works.

Limited accountability
: Passive leaders fail to hold themselves accountable for their roles in driving progress, hindering overall team accountability and performance.

Resistance to change
: Leaders' aversion to rocking the boat reinforces a culture of resistance to change, stifling creativity and progress.

To address these challenges, senior leadership has to step in to foster effective leadership by providing the following:

Clear expectations
: Define expectations for leaders' roles, responsibilities, and active involvement in driving improvements.

Open communication
: Establish communication channels with the concerned teams, facilitating regular feedback and discussions about needed changes.

Empowerment
: Empower leaders to make informed decisions by providing relevant information and insights to drive change effectively.

Innovation culture
: Cultivate a culture encouraging innovation, where leaders actively support and champion new ideas and approaches.

Accountability
: Hold leaders accountable for driving improvements and making decisions contributing to the team's growth and success.

In summary, the passive leadership antipattern hinders progress and stifles innovation through inaction and indecisiveness. Fostering a proactive and engaged leadership style that provides clear direction, gives actionable feedback, and embraces change is essential for creating a dynamic and thriving team environment.

UNDERAPPRECIATION

The *underappreciation antipattern* highlights a leadership misstep in which positive traits and behaviors exhibited by team members go unnoticed or underappreciated. This pattern emerges when leaders fail to readily acknowledge and celebrate commendable actions, such as bull's-eye commits, focused work, or self-motivated code cleanup. By disregarding these virtues, leaders inadvertently dampen team morale, hinder motivation, and miss opportunities to reinforce positive practices.

This antipattern results in various issues, but the most important is missed opportunities to present positive traits as examples for the rest of the team. Neglecting to recognize positive behaviors undermines the reinforcement of desired practices, hindering their consistent adoption. A culture that fails to appreciate positive traits can promote complacency and discourage efforts to go above and beyond.

Experiencing underappreciation from a manager can also have a psychological impact on employees, especially those who are silently doing their best to get things right. They might feel invisible, which would affect their motivation, morale, engagement, and overall well-being. Over time, employees may become demotivated, leading to decreased productivity and a sense of disconnection from their work.

It is essential that leaders foster a positive environment through the following practices:

Regular recognition
: Implement a practice of consistently acknowledging and celebrating positive traits, no matter how small.

Timely feedback
: Provide timely feedback to team members, recognizing their contributions immediately after they occur.

Public appreciation
: Publicly acknowledge and commend positive behaviors, contributing to a culture of recognition and encouragement.

In conclusion, the underappreciation antipattern can erode team morale and stifle the behaviors that drive progress and excellence. By adopting a practice of consistent and genuine recognition, leaders can cultivate an environment where positive traits are appreciated, team members are motivated, and the team's collective success is amplified.

Conclusion

Antipatterns serve as cautionary tales, shedding light on pitfalls that can lead to inefficiencies, misalignment, and overall project failure. By recognizing the signs of individual antipatterns, such as over-helping or trivial tweaking, teams can promote healthier collaboration and growth. Process-related antipatterns involving PR or retrospective irregularities underscore the necessity of structured methodologies, accountability, and well-defined workflows. Structural antipatterns highlight the significance of holistic team structures. Leadership antipatterns, like skeptical or passive leadership, remind us of the importance of active, empowering, and communicative leadership to steer projects toward success.

In navigating these diverse antipatterns, leaders are armed with the tools to assess their practices critically, identify areas for improvement, and cultivate a culture of continuous growth. Acknowledging and addressing these antipatterns head-on can pave the way for smoother collaboration, enhanced decision making, and long-term project goals. Although antipatterns explore the darker side of software engineering, the journey toward effective software engineering is illuminated by acknowledging that they exist and finding ways to beat them.

Navigating the various antipatterns in practice will be challenging, yet it provides opportunities for leaders to continuously learn and grow. Just as continuous improvement is fundamental to effective software engineering, it is equally crucial for effective leadership. Embracing these learning experiences is the transformative force that elevates ordinary leaders, enabling them to evolve into impactful and effective managers who can steer teams toward success with insight, resilience, and continuous growth.

6

Effective Managers

Throughout the preceding chapters, we have dived deeply into the intricacies of software engineering leadership, inspecting the essential qualities that research deems necessary for effective management and dissecting the steps required to tackle specific team issues. We have also explored the lurking antipatterns that leaders must identify and shield themselves and their teams from. Now, it's time to weave these threads together into a cohesive image of an effective manager.

In this chapter, we will focus specifically on the operational role of a manager, as distinct from the broader leadership aspects discussed in Chapter 7. Managers are responsible for maintaining order and consistency within established frameworks, which involves planning, organizing, and controlling resources and processes to ensure that day-to-day operations run smoothly, efficiently, and predictably.

Let's picture Casey—a seasoned engineer with a decade of experience in various engineering roles. Casey's profile has transformed from that of a junior to a senior software engineer and then to an engineering manager. Would Casey benefit from the knowledge of how to enable and expand effectiveness or recognize antipatterns in her team? Certainly, yes! Casey would gain foresight about antipatterns to anticipate and be better prepared to face them head-on. Nevertheless, she would still grapple with the day-to-day intricacies of transitioning into a managerial role and navigating the associated responsibilities with finesse.

You may not match Casey's profile exactly, but you may be contemplating a leap into the managerial track. Alternatively, you could already be a software engineering manager striving to refine your effectiveness in the role. Regardless of where you stand, I aim to provide valuable answers and actionable insights in the forthcoming sections.

In this chapter, we'll explore three critical areas managers should focus on. First, I'll discuss the mindset shift required when transitioning from individual contributor to people leader. Next, I'll provide tips for time management, people management, and project management to help managers juggle responsibilities more effectively. Finally, I'll offer strategies to foster team development, handle group dynamics, and proactively grow management skills through learning and networking. These insights will equip you with a deeper understanding of how to embark on this journey and, more importantly, how to be effective at it daily.

From Engineering to Management

Many engineering managers start their careers as engineers. In Chapter 3, I talked about how when you grow as a leader, it is more about people and less about your technical expertise. Essentially, going from being an engineer to a senior engineer and then an engineering manager requires a change in mindset.

Note

I felt like I was on the path to being a senior engineer at Google when I realized that, to scale myself, I had to shift my mindset from "me" to "we." By collaborating with others, sharing what I learned, and focusing on lifting the skills and expertise of people around me, "we" started to get so much more done.

Transitioning from a software engineering position to a management position or stepping into a management role for the first time can be challenging. While the exact challenges can vary based on the organization's size and structure, there are common difficulties that individuals often encounter. These difficulties often revolve around the shift from a technical, task-oriented role to one that emphasizes people management, communication, and strategic thinking.

In smaller organizations or startups, the shift may be subtle but challenging. In these environments, engineers may wear multiple hats and be deeply involved in the technical aspects of projects. When transitioning into a management role, they may need to overcome their urge for hands-on technical work and instead focus on guiding and supporting their team. For example, a senior software engineer who becomes a development team lead may initially struggle with not being directly involved in coding tasks they were previously responsible for. They must allocate tasks, mentor junior engineers, and oversee project timelines.

In larger organizations with established teams, there may be a different set of challenges. Here, management roles often require a heightened emphasis on communication, coordination, and strategic thinking. For instance, a software engineer transitioning to a mid-level management role in a larger tech company may need to navigate complex team dynamics, ensure cross-functional collaboration, and communicate project progress to various stakeholders. These responsibilities may demand more time in meetings, resolving conflicts, and aligning team goals rather than writing or reviewing code.

Regardless of organization size, new managers may feel more in control of a project by getting involved in every technical detail and design decision. As a result, they may fail to empower team members to solve problems independently. This micromanagement antipattern (discussed in Chapter 5) can make it challenging to pivot from an individual contributor mode, where you have more time for focused work, to spending more time on active team communication and coordination.

Prioritizing people management over technical expertise is easier said than done. While technical skills are valuable, a manager's success largely depends on their ability to lead and inspire a team. An engineer's strength or success comes from coding. However, engineers who become managers should consciously spend less time doing focused technical work, like coding. Instead, they must invest in one-on-one mentoring, running team meetings, working with different teams, and managing people.

The nature of the satisfaction you get from your work also changes in a managerial role. It can be challenging to let go of the immediate gratification of solving technical problems. You must wait to measure the impact of your actions as a manager. For example, finishing a coding task gives you immediate results. But, imagine you've just had a one-on-one with one of your team members and reassured them about some of their concerns. You have likely built trust in that interaction, which can motivate the employee. However, you must wait a while to determine if this motivation leads to positive outcomes.

As discussed in Chapter 4, developing team members, setting strategic directions, effective communication, and fostering a positive work environment become paramount in a managerial role. This shift in priorities requires a different mindset and skill set. Managers must clearly convey their vision, expectations, and goals to the team. They must also excel at active listening to understand team members' concerns and needs. Communication extends beyond the team, as managers often liaise between their teams and upper management or other departments.

Transitioning from an individual contributor role to a managerial role can be challenging, as it requires a significant shift in mindset and responsibilities. Let's consider the story of Priya, a skilled software engineer who recently stepped into the role of engineering manager.

Priya's initial days as a manager were overwhelming. She struggled with letting go of her coding responsibilities and found herself tempted to dive into the technical details of her team's work. However, she soon realized that her role was no longer to be the "star player" who solves every problem, but rather a coach who empowers her team to excel.

To build trust and foster open communication, Priya embraced transparency. She openly acknowledged her inexperience and assured her team that she was committed to learning and growing in her new role. She regularly shared project updates, goals, and challenges with her team, encouraging team members to voice their concerns and ideas.

Recognizing the importance of staying up to date with the latest technologies and industry trends, Priya dedicated time to attending workshops and participating in technical discussions. However, she understood that her primary value as a manager was in creating an environment that encouraged innovation and knowledge sharing. She organized brown-bag sessions and paired experienced engineers with junior team members to facilitate learning and collaboration.

By focusing on creating a psychologically safe environment, Priya enabled her team to have honest conversations and quickly adapt to changes. As she settled into her new role, Priya gained confidence in her ability to lead and support her team effectively.

Any transition similar to Priya's cannot be easy and comes with its own set of bumps and challenges. While technical expertise remains valuable, success in a management role hinges on leadership, effective communication, and navigating complex team dynamics and organizational structures. Adaptability, a willingness to learn, and a focus on personal growth are crucial for overcoming these challenges and thriving in a management role. Let's see how you can channel your strengths to overcome these challenges.

Getting Started

Regardless of your most recent background—whether you're an engineer moving into a management role or a seasoned manager—starting a new managerial role can be challenging, especially if you're joining an ongoing project. Feelings of imposter syndrome and being overwhelmed are common. During the first few weeks in a new managerial role, making a positive and productive start is crucial while addressing some of the immediate challenges that may arise. Here are some essential tasks that you can take up to ensure that you make a strong start:

Meetings with team members

Get to know your team members individually. Learn about their strengths, weaknesses, and concerns. Building rapport is essential. Schedule one-on-one meetings with each team member to introduce yourself, build rapport, and learn about their roles, responsibilities, and concerns. Use these meetings to listen actively, ask questions, and gather insights about the project and team dynamics.

Project assessment

Conduct a thorough evaluation of the project's status. Understand its objectives, scope, current situation, and any challenges it's facing. Review project documentation, timelines, and milestones to understand its current state. Identify any critical issues or bottlenecks that require immediate attention.

Understand the tech stack

Gaining familiarity with the project's tech stack will help you make informed decisions, provide relevant guidance, and empathize with the team's technical challenges. This will facilitate a smoother transition into problem-solving by helping you address technical intricacies.

Address immediate concerns

If there are pressing concerns or challenges within the project or team, take immediate action to address them. They could include things that affect the team in the long term, such as lack of clarity on project priorities and goals. Or they could include more mundane things, such as approvals for purchase of licensed software. In either case, try to get to the root of the problems, be open and transparent in your communication about these issues, and involve the team in problem-solving if possible. For example, you could facilitate a discussion with the team to address questions related to project goals or communication issues. When faced with issues related to approvals, you could identify the bottleneck through discussion and take the right steps to expedite the required approvals.

Identify quick wins

Look for opportunities to make immediate, positive changes or improvements within the project or team. Celebrate quick wins to boost team morale and demonstrate your commitment to progress.

Start networking
: Connect with key stakeholders inside and outside your team. Develop relationships with other department heads, if applicable. Networking and building alliances can be valuable in addressing larger organizational challenges.

Start prioritizing
: Identify critical tasks and priorities within the project. Focus on what needs immediate attention and what can be addressed later. Delegate responsibilities when possible, but stay involved in crucial decision-making processes.

Set up essential communication channels
: Establish clear communication channels and expectations with your team. For example, let team members know what is the best time to reach you and how. Also, elaborate on how soon they can expect a response from you for specific situations and when it is OK to tag certain communication as urgent. Decide on regular team meetings, updates, and reporting mechanisms.

Reflect and engage in self-care
: Take time for personal reflection and self-care. Starting a new role can be overwhelming, so maintain a healthy work-life balance to stay focused and energized and to prevent burnout. In my experience, burnout happens slowly and ends in apathy toward your work. Take care of yourself to stay motivated and energized in a new role.

Manage imposter syndrome
: Recognize that imposter syndrome is common among new managers. Understand that you were chosen for this role based on your skills and potential. Keep a record of your achievements and positive feedback. Remind yourself of your capabilities when self-doubt arises.

I realize this is a long list, but you cannot expect to start running at full speed on day one. You can start strong by doing all the things described, but trust and positive relationships aren't instantaneous; they're built over time. In your first week, it's essential to balance addressing immediate concerns and setting the groundwork for long-term success. As you navigate the challenges, maintain an open and adaptable approach to leadership and stay committed to continuous improvement.

Defining a Strategy

As you settle into your role, start working on an overall management strategy. Having a strategy helps you navigate some of the complexities of project management and leadership. Some important components that you must focus on when defining a strategy are as follows:

Long-term strategic vision
: While short-term goals are essential for immediate success, a long-term strategic vision is equally vital. Effective managers balance these by envisioning where the team and product should be in the future. This vision serves as a North Star, guiding your decision making and keeping the team motivated and aligned with a shared purpose. I have discussed in previous chapters how tools such as OKRs help to set and track objectives that align with your long-term vision.

Transparent tracking of objectives
: Transparency is a cornerstone of effective management. Keeping goals, roles, and progress visible not only fosters a culture of accountability but also ensures that everyone is aligned toward a common purpose. Invest in tools and practices that enable clear communication of objectives and progress tracking. Create, assign, and track project tasks and objectives using project management tools.

Data-driven decision making
: In the era of data abundance, leveraging metrics and analytics is paramount for rational and effective decision making. Your management strategy should include a commitment to utilizing data to understand trends, make evidence-based choices, and continuously improve team performance. Some project management tools may include a dashboard and reporting modules that can provide you with sufficient data for effective decision making.

Calculated risk management
: Effective management often requires knowing when to take calculated risks and when to be more conservative to stay ahead. Striking the right balance between innovation and stability is crucial. You must be able to assess potential benefits against the associated risks, making informed decisions that drive progress while mitigating potential setbacks. You can

also use techniques such as SWOT (strengths, weaknesses, opportunities, and threats) analysis to assess a decision and calculate the risks.

As you craft and refine your management strategy, remember that it's a dynamic process, one that evolves as your experience and understanding grow.

Managing Your Time

Managing time as a software engineering manager is tough because you may have to guide the team in handling a few technical challenges besides team management. Just as you try to do some deep work (*https://oreil.ly/-KFl7*) (work you do over a prolonged period when you can concentrate on a difficult task), you get interrupted by team members, other departments, or your boss. You frequently have to deal with unforeseen project challenges and may need to make tough decisions requiring in-depth analysis of situations.

Communication, although necessary, may take away a lot of your time. You may feel that you are always talking or communicating; meetings, emails, calls—it all takes time and can make you lose track of what you are doing. As a manager, you must spend a lot of time communicating to manage your team, help team members, and resolve conflicts. But where can you find all that time?

Through all of this, it might become challenging to keep yourself and your projects on track. You might find it hard to focus on any planned work for long periods, which might lead to feelings of falling behind.

Utilizing a few time-tested time management techniques is essential to preventing you from getting overwhelmed. Let me recommend a few to help you plan your time management strategy and then execute and assess it.

PLANNING

Although it may sound obvious, come up with a plan for how you use your time. By setting boundaries around your time, you allow yourself to concentrate on tasks and produce better results. The following are some ways you can proactively plan to manage your time:

Time blocking

Time blocking is a time management technique in which you schedule specific blocks of time for different tasks or activities in your day, helping you stay organized and focused. It's like setting appointments with yourself for work, meetings, or other responsibilities to improve productivity and manage your time effectively. Use time blocking to assign specific periods for focused work, team meetings, and administrative tasks. Decide to protect these blocks from interruptions.

Chunking similar tasks

Group similar tasks together when possible. For instance, handle all financial reviews or project planning in a single dedicated block of time. Context-switching between diverse tasks can be time-consuming and mentally taxing.

Planning your communication

Set specific times during the day to check and respond to emails rather than being continuously reactive. Use email filters and labels to prioritize and categorize messages, focusing on high-priority communications first. Create designated "office hours" when team members can approach you with questions or concerns. This consolidates ad hoc discussions. If you are using instant messengers for office communication, be wise about it. If you make immediately responding to all messages a habit, messengers can take over and eat up all your time.

EXECUTION

As a manager, you are likely not going to be able to do everything that you want or planned for. However, it's important to achieve the most important goals. Here are some practical steps that can help you in executing your plan.

Develop team members

Invest in coaching and mentoring your team members to handle specific tasks independently. They can take on more responsibilities as they grow, reducing your workload. Encourage skill development and cross-training within the team to ensure everyone is equipped to handle various tasks. This will also help inculcate autonomy, mastery, and trust in the team—essential motivators, as discussed in Chapter 1.

Delegation

Identify tasks that can be delegated to team members. Trust your team's capabilities and empower members to take on responsibilities. Delegate tasks based on team members' strengths and developmental goals, allowing them to grow and contribute more effectively. For example, you may delegate most of the document and code reviews to senior team members while conducting spot checks to ensure they meet organizational standards.

Learn to say "no"

Be selective about taking on new commitments or projects. Avoid overloading your schedule with tasks that don't align with your priorities. Politely decline tasks or delegate them when necessary.

ASSESSMENT

Finally, it is essential that you objectively verify whether your strategies are working through regular assessment. You must assess the execution of your time management strategies and adjust them if necessary as a cycle of improvement:

Calendar audits

Regularly audit your calendar to assess how you allocate your time. Consider weekly audits to evaluate how you're spending your time and whether it matches your priorities. Additionally, monthly reviews can help you identify recurring time drains that you can reduce. Ensure that your schedule aligns with your strategic priorities. If something's not in alignment, then reduce the time you spend on it. Assess whether the task can be delegated or declined (as discussed in the "Execution" section).

Reflect and adjust

Set aside time weekly or monthly for reflection. Analyze how you've been managing your time and adjust your strategies as needed. You can ask yourself questions such as these:

- Have I minimized distractions and interruptions during focused work sessions?
- Did my delegation strategy work as expected? Was I required to provide any assistance on the delegated work?
- What amount of time did I invest in personal development and self-care?

Be adaptable and willing to modify your time management techniques based on changing circumstances. Seek feedback from your team and peers to gain insights into your time management effectiveness.

These powerful techniques can help you effectively juggle leadership responsibilities, meet your deadlines, maintain a healthy work-life balance, and contribute to your team's success.

Understanding and Setting Expectations

In the preceding chapters, I have discussed how everyone should be clear about their role in an effective team. This applies to managers, too. As a manager, it's crucial to have a clear understanding of your own role and responsibilities. This clarity enables you to allocate your time and resources effectively, ensuring that you can focus more on the most critical tasks.

You can then elaborate on the key components of these expectations for your team members. You can create a framework that guides their efforts toward shared goals and success by defining what they are accountable for and what they should be working toward.

WHAT RESULTS ARE EXPECTED FROM ME?

It's essential first to understand the expected outcomes to become effective. Managers should understand the expectations placed on themselves as individuals and also on the teams under their leadership. Expectations can come from various sources, including your superiors, the organization's objectives, and the nature of your role. Some critical factors that can help you gauge these are as follows:

Communication
: Regular and open communication with your superiors and peers is essential. Discuss your role, responsibilities, and performance expectations. Seek clarification if needed and ensure alignment with organizational goals.

Goal setting
: Your organization may already use a goal-setting framework (e.g., OKR or SMART, as discussed in earlier chapters). Collaboratively set specific, measurable, and achievable goals with your superiors. These goals should align with the organization's objectives and provide a clear direction for your work.

Prioritization
: Prioritize your tasks based on their importance and alignment with organizational and project goals. Allocate time and resources accordingly to ensure that high-priority tasks are addressed first.

Self-assessment
: Periodically evaluate your own performance against the expectations set for your role. This could involve reflecting on your achievements and areas for improvement and adjusting strategies as needed to meet or exceed expectations.

WHAT RESULTS DO I EXPECT FROM TEAM MEMBERS?

Equally important is setting clear expectations for your team members. When team members understand what is expected of them, it fosters accountability and productivity. Here's how to do it effectively:

Clear communication
: Communicate the goals and objectives of the team and how each team member contributes to these goals. Use simple language and provide examples to ensure understanding.

Individual meetings
: Hold one-on-one meetings with team members to discuss their roles, responsibilities, and performance expectations. Encourage questions and feedback to ensure alignment.

Goal alignment
: Ensure that individual goals align with project and organizational goals. As discovered by Project Aristotle, impact is an essential factor affecting team members' motivation and effectiveness. When team members see how their work contributes to the bigger picture, they are more motivated and accountable.

Document expectations
: Maintain clear documentation of each team member's roles, responsibilities, and performance expectations. Your organization may already have a framework for this. This documentation can serve as a reference point during evaluations and discussions.

Effective management involves meeting and setting clear expectations. You should actively communicate with superiors to understand what they expect from a person in your position and use this understanding to allocate time and resources effectively. Simultaneously, you should set transparent expectations for team members, fostering a culture of accountability and ensuring each member knows their role in achieving shared goals. Two-way communication and alignment are essential for successful leadership and team performance.

Communication Essentials

There are multiple avenues to connect with team members. These include team meetings, one-on-one discussions, messenger platforms, emails, review comments on documents or code, and management tools like Jira. Furthermore, nonverbal communication, such as body language, plays a role. It may be challenging to recall every detail from each of these interactions, but it's essential that you maintain consistency in the overall principles and ideology you convey across these various channels. For instance, your team members should not be perplexed because you said one thing during a team meeting and assigned them a completely different task in Jira.

When a manager's communication lacks consistency, it can create a host of challenges for the team. Conflicting information or instructions may confuse the team members and result in uncertainty about their tasks and priorities. This can lead to wasted time and effort. If it happens too often, team members may become skeptical or hesitant to rely on the information provided. This affects team morale, motivation, and productivity in the long run.

To ensure consistent communication, you must have a clear management strategy that serves as your guiding principle in all interactions with your team. For instance, if you've established at the outset of a project that all API documentation must be reviewed by the product owner before publication, it's essential to consistently uphold this standard. If someone asks you for an exception by directly communicating with you, then either do not allow that shortcut or justify why you allowed it in a team setting.

Each method of communication is different. Some is written or verbal, some is virtual or in person, and some is synchronous or asynchronous. Despite that, you can still maintain consistency in your communication. The following are ways you can be effective and consistent with your communication over various channels.

TEAM MEETINGS

Meetings are necessary to set the course of the project, help individuals collaborate, and maintain momentum throughout. The frequency with which you should all meet as a team depends on the overall duration planned for the project.

On short projects (say a couple of months), meetings should not happen so often that they waste time. At the same time, the interval between two meetings should not be so long as to result in miscommunication. If different groups of people don't have opportunities to collaborate outside of team meetings, then a weekly meeting or check-in is essential.

On long or complex projects, it may be required that you meet more often as a team at the beginning, during the planning stage, and at the end, during the integration stage. The meetings may be less frequent in between, during the development stage.

In addition to the frequency of meetings, a few key things to note about team meetings are as follows:

Set clear meeting objectives
: Determine the agenda topics in advance, what decisions must be made, and what action items should result from the meeting. Communicate these objectives to your team in advance so that everyone comes prepared with relevant information and can contribute meaningfully to the discussion.

Keep meetings focused and time-bound
: Start and end meetings on time to respect your team members' schedules. This demonstrates your commitment to efficiency. Maintain a clear agenda and stick to it. Avoid going off-topic or delving into unrelated discussions. If new issues arise during the meeting, consider scheduling a separate conversation for them with just the key members affected by them.

Promote active participation and inclusivity
: Avoid monopolizing discussions and allow team members to speak. Foster an environment where all team members feel comfortable sharing their ideas, questions, and concerns. Encourage active participation, constructive contributions, and open dialogue, especially from remote team members. Rotate meeting roles, such as facilitator or notetaker, to involve different team members and distribute responsibilities.

ONE-ON-ONES

Conducting effective one-on-one meetings with your team members is a vital aspect of your role as a manager. The primary focus should always be on your team members and their needs. These meetings are a platform for team members to express themselves, share concerns, and seek guidance. Thus, active listening is paramount. When your team members speak, listen attentively, not just to the words they say but also to the emotions and underlying messages.

While listening, it's essential to refrain from immediately providing solutions to their problems. Instead, aim to empower them by guiding and coaching them when necessary. Encourage them to think critically and find solutions independently whenever possible. This not only promotes their problem-solving skills but also boosts their confidence in handling challenges. When you inquire and ask open-ended questions, it invites deeper discussions and prompts your team members to reflect on their own situations. Moreover, inquire about their goals and aspirations, not just their current tasks. This shows that you value their growth and development, reinforcing their sense of purpose and motivation.

Any feedback you provide during these sessions should be helpful. Feedback can be useless if it's not actionable (e.g., vague feedback that fails to address performance concerns). Be specific in your feedback, both positive and constructive, to help team members understand their strengths and areas for improvement.

When speaking, speak with a calm and composed tone, even during challenging conversations. Avoid raising your voice unnecessarily, as it can come across as confrontational. Instead, modulate your tone to reflect empathy, encouragement, or assertiveness as needed.

Additionally, make it a point to ask your team members how you can assist them in overcoming any obstacles they're facing. This collaborative approach demonstrates your commitment to their success and well-being. By collectively exploring strategies to address roadblocks, you not only strengthen your relationship with your team but also facilitate a culture of shared responsibility and support. In these one-on-one meetings, your role evolves from manager to mentor, guiding your team members toward greater autonomy and success.

In Chapter 4, I shared Google's template for effective one-on-ones with questions that cover most of the aforementioned points. Tailor this template as required and share it with your team members, too, so that they have an idea of what they can and should discuss.

MESSAGING

Effective messaging across various platforms is crucial for maintaining clear communication, managing tasks, and ensuring that the team remains aligned and focused. You are spoiled for options when it comes to communicating with your team from your desk or phone. Many organizations use emails, messaging apps, and task management software almost interchangeably. While all these modes help to maintain a record of the conversation for posterity, the key is to use each mode of communication strategically based on the nature of the message and the urgency of the situation. Each communication mode serves specific purposes, and to use them efficiently, you must know when and how to use each:

Email
: Email is suitable for formal communication and documentation. Use it for sharing important project updates, reports, or detailed information that requires careful consideration. Email is also a great choice when communicating with stakeholders outside the immediate team, as it allows for a more structured and permanent record of the conversation. However, it's not ideal for urgent matters, as it's typically not as real-time as other messaging platforms.

Instant messaging
: Instant messaging platforms such as Slack are excellent for quick questions, casual conversations, and real-time collaboration. Use instant messaging for day-to-day interactions with your team, such as sharing brief updates, asking questions, or providing immediate feedback. It's particularly useful for urgent matters that require a quick response. However, be mindful not to overuse it, as constant interruptions can hinder productivity. Set specific hours or designated "focus time" to avoid distractions from instant messaging.

Task management software
: Task management software is ideal for tracking tasks, assigning responsibilities, and monitoring project progress. Use these platforms to organize and prioritize work, assign tasks, and set deadlines. Regularly check project management software to ensure that everyone is on track and aligned with project goals. These platforms are less suitable for real-time communication but are invaluable for maintaining focus and organization.

Keeping track of all this communication can get overwhelming. It's essential to strike a balance in terms of the frequency with which you access these. Check email periodically throughout the day to ensure you don't miss critical updates. For instant messaging, establish specific time slots for checking messages, such as in the morning, after lunch, and before the end of the day. For project management software, incorporate regular check-ins into your workflow, such as an end-of-the-day review, to keep the team focused on tasks and goals.

It's also important to note that some communication is just not suitable for sharing via email or messenger. It should be shared face-to-face in person or via video meetings. For example, provide critical feedback privately in one-on-one sessions rather than publicly over messaging platforms. Similarly, if changes to the team structure are expected, informing team members in a brief meeting is better before sending a detailed email announcement.

Appropriate use of each messaging channel will help you organize your communication effectively without losing focus on your other tasks.

NONVERBAL COMMUNICATION

Effective nonverbal communication is a powerful tool for managers in conveying messages, building rapport, and fostering positive relationships within a team:

Body language
: Your body language can speak volumes. Maintaining an open and welcoming posture during conversations is important and conveys that you are receptive and approachable. Demonstrate attentiveness and engaged listening during conversations through body language. Lean slightly forward, maintain eye contact without staring, and nod occasionally as team members speak to acknowledge that you are listening. Use gestures to emphasize key points, but be mindful not to overdo it, as excessive gestures can be distracting.

Facial expressions
: Facial expressions are highly effective at conveying a wide range of emotions. Smile genuinely to create a warm and inviting atmosphere. When appropriate, mirror the feelings of your team members to show empathy and understanding. However, be authentic in your expressions; forced or insincere smiles can be easily detected.

Proximity
: Your physical proximity to team members can send subtle messages. Approachability is often associated with closer proximity, while standing too far away can create a sense of detachment. Be aware of cultural norms regarding personal space, as these can vary, and respect individual preferences.

Nonverbal communication is also important in remote settings. With the absence of physical presence, videoconferencing, emojis in written communication, and vocal tone take on added significance. Managers should pay special attention to these cues to ensure that their remote team members feel connected, understood, and supported.

People Management

People management is often a key component of your duties as an engineering manager. I have already covered periodic tasks such as conducting one-on-ones and task assignments in this chapter. While I have already shared some essential traits to consider when hiring effective engineers for your team in Chapter 1, I would like to touch upon certain macro aspects of hiring, performance appraisals, and attrition in this section.

Simply put, managing people effectively in the software engineering industry poses unique challenges and, depending on the mix of people you have, can become more complex than managing the technical side of the project. In Chapter 5, I mentioned some antipatterns that can surface because of the unique characteristics of different types of engineers, whose traits (even positive ones) can negatively impact a project. Some additional challenges that may affect you when managing people include the following:

Tech talent competition
: Attracting and retaining experienced tech talent within your budget is a constant challenge. The industry is highly competitive, with a shortage of skilled professionals.

Skill diversification
: The rapid evolution of technology requires team members to upskill and adapt to new tools and frameworks constantly.

Remote work dynamics
: With the rise of remote work, managers must effectively lead and motivate teams across different geographical locations and time zones.

Expectations
: Due to the competitive market, good software engineers expect continuous growth in terms of remuneration, perks, career prospects, and work-life balance.

Project complexity
: Managing multifaceted software projects with cross-functional and hybrid teams to meet deadlines can be intricate and demanding.

Here are a few areas where you might need to use your people management skills.

HIRING

The hiring process is critical. Before you share your requirements with human resources, clearly define job roles and identify essential skills for which you are hiring. Employ structured interview processes to assess technical proficiency, problem-solving abilities, and cultural fit. Sometimes, you may need to rush to fill vacancies to meet project timelines. As far as possible, try to prioritize attitude over skills because someone who cannot work as a good team player may have a disruptive influence on the team.

We have discussed the traits that contribute to an effective engineering mindset in Chapter 1, and we summarize them in Figure 6-1. Look for engineers who are able to discuss how they've demonstrated these traits in the past.

Figure 6-1. Ten key traits of effective engineers

PERFORMANCE EVALUATION

Regular performance evaluations of team members should focus on both technical skills and soft skills like communication and teamwork. Medium-to-large companies usually have well-defined performance processes and norms that you will follow. You would most likely use the company evaluation form or template. Regardless of your organization's established performance evaluation process, which may not reflect much beyond the employee's performance metrics, you should make a point to discuss their personal career aspirations. Here are some important areas you need to address with the team member and specific questions to ask the team member during the performance discussions:

Career goals
: What are your short-term and long-term career goals, and how can I help you achieve them within the team or organization?

Professional development
: What new skills or technologies would you like to learn or improve upon in the coming year? Are there any training or development opportunities you believe would benefit you and the team?

Additional responsibilities
: Do you have any interest in mentorship or leadership roles within the team? Would you be interested in customer-facing opportunities with greater interaction with end users, which may require you to travel or represent the team on social media and in conferences?

Challenges and achievements
: What were the most challenging tasks you worked on during the evaluation period, how did you overcome them, and what did you learn from the experience?

Feedback and improvement
: What feedback do you have for the team or myself (your manager) on improving work processes, tools, or team dynamics? How can I support you in achieving even better results in the next evaluation period?

Work-life balance
: Are you able to manage your time and work effectively? Have you experienced any work-related stress or difficulties with balancing your work and personal life?

These topics and questions can provide a more holistic view of a software engineer's performance and career aspirations, facilitating a more constructive and personalized performance evaluation discussion.

Your role in the discussion is to listen attentively and provide meaningful and constructive feedback where necessary. Based on their response to the questions, discuss and individualize development plans to help team members progress in their careers.

In addition to the formal performance evaluation discussions, I check in with my team regularly to share high-level plans for the coming quarter or year, discuss achievements in the past year, and provide feedback. This also gives team members an opportunity to share their aspirations or expectations in the coming year or quarter.

ATTRITION MANAGEMENT

Even if you have a healthy work culture, effective teams, and opportunities for growth, some people will choose to leave for various reasons. For example, they may leave for better career opportunities, retire, or be unhappy with the work culture of your organization. You can try to retain talent by offering competitive compensation, opportunities for skill growth, and a positive work culture. You can also regularly check in with team members to understand their concerns and address any burnout proactively. However, attrition can still take place for reasons beyond your control. When this happens, it's essential to handle it gracefully. A playlist for such situations would be to:

1. Start by setting up a meeting to understand the reasons behind the departure. Reasons for leaving can be personal or professional.
2. If their reasons are related to the team or yourself, solicit feedback. Ask about their experience on projects and with work relationships. Identify areas that may need improvement.
3. If they are leaving due to a personal situation, offer hybrid or flexible work options if it will help you retain a dedicated employee without significantly affecting the project they are working on.
4. Use feedback from exit interviews to identify areas for improvement, if any, within the team or organization.
5. Ensure a smooth transition of knowledge and responsibilities to prevent disruption of ongoing projects.

In addition to exit interviews with the person who is leaving, consider conducting stay interviews (*https://oreil.ly/b9y1F*) with current team members to proactively address concerns and reduce attrition due to the same reasons in the future. You may already discuss issues related to employee engagement and satisfaction in one-on-ones or performance appraisals. Additionally, stay interviews provide an opportunity to gain deeper insights by allowing you to ask team members targeted questions such as these:

What aspects of your job do you find most fulfilling or least enjoyable?
: This question encourages the team member to consider the most positive and negative aspects of their role in the team and discuss them.

Are there any unmet career or development needs you'd like to discuss?
: While you may have discussed future aspirations earlier, this question allows you to uncover missed opportunities for growth and development.

If you could change something about your job, what would that be?
: This question encourages team members to talk about their vision for their role.

These questions aim to uncover deeper insights into team members' motivations, career aspirations, and the factors that contribute to their job satisfaction. They can guide managers in creating a more engaging and supportive work environment.

While most employees leave by themselves, sometimes organizational changes or constraints may necessitate letting go of team members. Approach this with empathy, transparency, and fairness. Communicate the reasons clearly and provide support during the transition, such as outplacement services or referrals.

MENTORSHIP AND COACHING

Mentorship and coaching have proved to be extremely handy tools for developing talent. *Mentorship* is a long-term relationship between a more experienced person (mentor) and a less experienced person (mentee) in which the mentor provides guidance, support, and encouragement to the mentee. *Coaching* is a more focused and short-term relationship between a coach and a coachee in which the coach helps the coachee identify and achieve specific goals.

Both mentoring and coaching can help engineers learn from experienced colleagues or external experts, accelerating their acquisition of new skills and knowledge. This can range from mastering programming languages and frameworks to developing soft skills like communication, collaboration, and problem-solving.

A successful mentoring program in a tech team could start by identifying mentees through methods like performance reviews or colleague nominations.

Specific goals might be set, such as database or performance optimization techniques in specific project-related contexts. Regular meetings should be scheduled for discussing progress and challenges and for setting short-term objectives. The mentor could provide resources, such as recommended readings or relevant assignments to work on, to facilitate the mentee's development.

Invest in mentorship programs and coaching to nurture talent from within your team. Such programs allow you to provide guidance, career advice, and educational opportunities, which can boost team morale and personal growth. Empower senior engineers to mentor junior colleagues, fostering a culture of continuous learning. Consider rotational assignments like having senior engineers mentor newer members on projects. Establish formal mentorship programs to foster skill development and relationship building.

While mentoring is a growth opportunity for team members, it is also beneficial to management. When the time comes, you have someone ready to take up a senior role within the organization and can avoid hiring someone from outside. Here are some key benefits of training up your existing talent (by providing mentorship/coaching programs) over hiring outside talent:

Cultural fit and alignment
: Team members are already familiar with the company culture, values, and existing dynamics, which leads to quicker alignment and collaboration.

Cost and time savings
: Promoting from within reduces recruitment and onboarding costs and shortens the time to productivity, making it a cost-effective choice.

Employee development and morale
: Mentoring and coaching demonstrate a commitment to employee development, boosting morale, motivation, and overall job satisfaction.

Effective people management in software engineering requires a blend of emotional intelligence, communication skills, and adaptability. With active involvement and careful consideration when hiring, appraising, and addressing concerns, you can foster a culture of growth, empathy, and collaboration.

Managing Challenging Projects

There may come a time or several times in your career when you are faced with challenging projects that don't go smoothly. They may make you feel like you're on a roller-coaster ride, with many twists and turns and highs and lows.

Although challenging, these projects are great learning opportunities that can teach you to adopt a multifaceted approach to project management.

Emily, a friend and a seasoned software engineering manager, once found herself thrust into a challenging project with a history of setbacks. The project, which involved integrating multiple systems and technologies, had a reputation for its complexity. As Emily delved into the project's details, she quickly realized the magnitude of the task at hand.

The project had a history of sudden changes in requirements and specifications, making it difficult to maintain a cohesive plan. Additionally, the team had faced numerous technical hurdles in the past, resulting in delays and frustration. Emily knew that she needed to take decisive action to rein in the project and set it on the path to success.

Drawing on her experience and leadership skills, Emily began by rallying her team and instilling a sense of purpose and determination. She emphasized the importance of clear communication and collaboration, encouraging team members to voice their concerns and ideas openly. Despite the project's tumultuous history, Emily remained undeterred, and she instilled the same confidence in her team. Through her strategic guidance and unwavering commitment, Emily led her team through the challenges and uncertainties.

Emily's leadership ultimately proved instrumental in the project's successful completion, and through her journey, she experienced a profound learning curve that enriched her leadership skills and insights. She discovered the importance of adaptability and resilience in the face of adversity, as well as the value of open communication and collaboration in fostering team cohesion. The project provided Emily with invaluable lessons in navigating complexity, managing uncertainty, and driving progress in dynamic environments. Emily and her team emerged from the project with a newfound sense of capability, armed with the knowledge and skills to tackle future challenges with poise and determination.

Unexpected challenges and opportunities demand that your team adapts continuously with you in the driver's seat.

Here are a few key considerations that can help you navigate through any roller-coaster ride of a project:

Agile approach

Consider embracing an agile project management approach that allows for flexibility in scope, resources, and timelines. Agile methodologies like Scrum or Kanban enable you to respond swiftly to changes and adapt to new information as the project progresses. As mentioned earlier in

Chapter 1, ensure that you tailor the methodologies and set up tools and frameworks based on your needs.

Scope management
: Ideally, you will define the project's objectives and scope at the outset. However, once you have experienced a few twists—be prepared for more. Regularly review and adjust the scope as necessary to accommodate new requirements or unexpected challenges. Use techniques like user stories and backlog prioritization to keep the focus on delivering high-value features. Not all problems are of equal significance. Prioritize the issues based on their severity and their potential to derail the project. Triage them to address the most critical problems first.

Prototype
: Quite a few of such projects include business requirements, where your team might need to use a new technical concept, product feature, or interface that has never been used before. Before committing to anything, ensure your team has prototyped the possible solution or developed a proof of concept. This proactive approach helps identify and mitigate potential technical hurdles but also provides an opportunity to refine project plans, allocate resources more efficiently, and ensure that the project remains on course.

Decisive but flexible
: Make decisions promptly but be ready to adapt as situations evolve. Ensure that your team is well equipped and motivated to handle the project's ups and downs. Address skill gaps as they arise. Modify the project plan, timeline, and resource allocation as needed to accommodate new circumstances. Consider revising scope, adjusting deadlines, or reallocating resources to address the problems effectively. Understand that time constraints may fluctuate as the project unfolds. Discuss and communicate schedule adjustments transparently with stakeholders, management, and team members.

Quality control
: While upholding quality standards in demanding projects can prove challenging, neglecting to do so will inevitably compound the issues at hand. Rigorous code reviews and testing must be performed at all times. Constructive code reviews engage team members to identify and rectify issues early. Complemented by various testing methods, such as unit, integration, and system testing, the team can ensure that bugs and inconsistencies

are identified and addressed promptly. Automated testing and continuous integration can also help in ensuring that changes don't introduce new issues.

Work-life balance
: Recognize the importance of work-life balance for yourself and your team members. Encourage breaks and time off to prevent burnout. Promote a culture of self-care and support to ensure that team members can sustain their energy and creativity throughout the project. Encourage delegation and sharing of responsibilities so that every team member, including yourself, gets some time to rest and zone out.

Communication
: Maintain open and transparent communication channels within the team and with stakeholders. Regularly hold stand-up meetings, retrospectives, and review sessions to discuss progress, challenges, and adjustments. Keep stakeholders informed about project developments and changes to manage expectations effectively.

Removing blockers
: Be proactive in identifying and eliminating project blockers. Encourage team members to report obstacles and bottlenecks promptly. Collaborate with other teams to address issues that impact project progress.

Celebrating successes
: Amid the challenges, you may not have the time or energy to do so, but remember to celebrate successes, no matter how small. Recognize the team's achievements to boost morale and motivation.

Managing roller-coaster-like projects is about embracing change, staying adaptable, and maintaining clear communication. By adopting an agile mindset and addressing these key aspects of project management, you can lead your team to success, even in the most unpredictable project environments.

Managing Team Dynamics

Although you can hire great people, you can't always predict how they will behave or how they will interact with others. Humans and human interactions are complex, and problems are likely to surface as time progresses. We have already discussed a few individual and team collaboration antipatterns in Chapter 5. In an era of remote work and global collaboration, these challenges can multiply.

INDIVIDUAL IDIOSYNCRASIES AND DIVERSE TEAMS

Individuals differ in a number of ways. In a diverse group of people, you will have introverts and extroverts, and you will also have people with different skills/educational backgrounds, problem-solving skills, work ethics, cultural/economic backgrounds, and genders. Their biases and perspectives will differ based on how they got configured over time.

To overcome any issues arising from these differences, consider organizing team-building activities to help people share their views in an informal setting and learn to understand and respect others' opinions.

You can also assign tasks that align with each team member's strengths, weaknesses, or preferences. For instance, introverts might excel in in-depth research, while extroverts can shine in client meetings. Take the time to understand your team members and delegate tasks based on their capabilities.

Additionally, promote open dialogue and encourage team members to voice their preferences. Be open to new ideas and try cultivating a culture where team members feel their contributions are valued. Once team members get comfortable in the team and start trusting each other, you might ask them to diversify to help them acquire new skills and grow.

REMOTE TEAMS

Asynchronous communication can hinder collaboration in a remote team spread across different time zones. To address this, managers can establish clear communication protocols, including preferred channels and response times. Encourage the use of collaboration tools that allow for real-time updates, file sharing, and videoconferencing to bridge the geographical gap.

Remote team members may struggle with feelings of isolation and the challenge of disconnecting from work. Managers should emphasize work-life balance and set expectations regarding working hours. Regular virtual check-ins and in-person team-building activities, such as company retreats for fully remote teams, can help combat isolation and foster a sense of belonging.

CONFLICT RESOLUTION

Recognize that conflicts often arise from differences in personalities, work styles, or interpretations of goals. For example, you may have two very different team members, one who prefers a highly structured and organized approach and another who is more inclined toward a flexible and creative work style. As they collaborate on the project, conflicts arise due to these differences.

To resolve conflicts effectively, you must be able to navigate interpersonal issues within your team. Instead of letting these disputes linger, address them promptly by creating an open space for dialogue where team members feel comfortable discussing their concerns and differences. This not only helps you understand the root causes of conflicts but may also allow you to create a win-win situation. In the preceding example, the individual with the creative work style may contribute innovative solutions to problems while the more organized individual can ensure that the solution accommodates all the requirements and constraints of the project.

Be alert and don't allow conflicts to derail your team's progress. Maintain a strong focus on your team's collective mission. Regularly remind your team of these common goals to minimize conflicts related to personal agendas. Through adept conflict resolution, you can turn disputes into opportunities for growth and team cohesion.

Enabling Mastery and Growth

In Chapter 1, I shared how a sense of progress toward mastery is a powerful motivator. Believing that they are growing and improving their capabilities contributes to team members' inner drive. Yet enabling growth can be challenging, especially at times when the team is already continuously busy on a challenging project and has no time to try something new. Even when the situation is comparatively relaxed, team members may not be sure which skill would add more value to their profiles. As a manager, your role in facilitating growth remains crucial regardless of the team's operational state.

HARNESSING DOWNTIME FOR GROWTH

During slow periods, it's essential to invest in individual growth.

Many talented engineers may be idle for some time when they are between projects. If you are managing such a group, you can guide them through brainstorming sessions where they come up with ideas for mini-projects. Your team could invest the time in building tools that would be useful to them in a future development cycle. The requirements for such tools could be based on issues identified in a previous cycle. The mini-project could aim to address these issues to improve the developer experience and productivity.

Additionally, you can engage team members in conversations about their career goals and interests and help them come up with a learning plan to align with their aspirations. You can also foster mentorship and peer learning within the team. Experienced members can guide others, and peer-to-peer knowledge-sharing sessions can be highly effective.

If team members express interest in a specific training or certification course, encourage them to pursue it. Your organization may also approve sponsorship for short courses; if there are a good number of trainees available, such courses can be conducted in-house.

EMPOWERING GROWTH AMID HIGH-WORKLOAD PERIODS

Even during hectic periods, growth should not be relegated to the background. Encourage learning through on-the-job experiences by assigning tasks that stretch abilities and provide opportunities for growth within ongoing projects. Provide constructive feedback and recognize growth efforts, motivating individuals to persist in their learning journeys.

Code review exercises are great avenues for senior engineers to share knowledge tidbits with others. You can similarly embed learning into other daily work processes. Foster a culture of peer support by promoting informal mentoring interactions, allowing team members to seek advice and guidance from colleagues.

Help team members manage their time efficiently, setting aside dedicated intervals for self-improvement, even if it's only for 10 to 15 minutes a day.

In both downtime and busy periods, nurturing growth requires a proactive approach. By tailoring growth strategies to team members' unique circumstances and individual aspirations, you not only enable their development but also contribute to their motivation and overall job satisfaction.

Networking Essentials

I have talked extensively about collaboration among team members in the previous chapters. As a manager, you will be required to collaborate with various people within and outside your organization—stakeholders, colleagues, industry partners, and many more. Networking with these people on a regular basis rather than reaching out to them only when you need something is essential.

Networking is not just about building a list of contacts but forging meaningful connections that can pave the path to elevating your management skills and enhancing your team's performance. Here are some of the ways networking can help you:

Knowledge exchange
: Networking allows you to tap into a wealth of knowledge and diverse perspectives. It's a dynamic way to stay informed about industry trends, emerging technologies, and evolving management practices.

Problem-solving
: Networking provides a pool of resources to help you tackle complex problems. You can seek advice from experienced peers, bounce ideas off colleagues, or collaborate with external experts who have encountered similar issues.

Professional growth
: Networking fosters your professional development. By attending industry events, seminars, and conferences, you can stay updated on the latest developments that can be beneficial for yourself or your team.

Collaboration opportunities
: Effective management often requires collaboration among departments or organizations. You can cultivate a network of trusted connections who are more likely to engage in collaborative efforts. These partnerships can lead to joint projects, shared resources, and mutually beneficial initiatives.

Networking may come naturally to some, while it may be more difficult for others. Regardless, here are some things to remember to network effectively:

Be genuine
: Approach networking with authenticity and a genuine interest in others. Authentic connections are more likely to yield fruitful relationships.

Listen actively
: Actively listen during conversations. Understand others' perspectives and offer valuable insights when appropriate.

Follow up
: After initial interactions, follow up with your contacts. This could be a simple thank-you note or an email sharing more information on a topic you discussed.

Maintain regular contact

Networking isn't a one-time effort. Stay in touch with your connections regularly, even if it's just to catch up over coffee or attend industry events together.

Diversify your network

Don't limit your network to your immediate field. Connect with individuals from diverse backgrounds and industries to broaden your horizons.

Aim to engage in networking events or activities at least quarterly. Attend industry conferences, join professional associations, and participate in webinars or meetups relevant to your field. Additionally, set aside time each month for coffee meetings or virtual catch-ups with your network contacts to maintain and nurture these relationships.

Remember that effective networking is not just a valuable skill: it's a strategy that can drive your success and contribute to your team's growth and achievement of organizational goals.

Conclusion

The previous chapters provided a research-backed framework for effective leadership, encompassing a spectrum of key principles. In this chapter, I've aimed to bridge the gap between these theoretical ideas and the practical execution of essential managerial processes. Here are a few key ideas that we discussed:

- By adopting time management techniques, software engineering managers can efficiently balance the use of their technical expertise with leadership responsibilities.
- Consistent and appropriate communication, whether through well-structured team meetings, one-on-one discussions, or messaging platforms, is pivotal for keeping the team aligned and motivated.
- Networking outside the immediate team provides valuable insights and opens doors to collaboration.
- People management, covering hiring, performance appraisals, and, when necessary, the challenging process of letting team members go, plays a pivotal role in building a cohesive and high-performing team.
- In the face of challenging projects, a proactive approach helps navigate uncertainties and keeps the team on track.

Overall, managerial processes intertwine to create a web of responsibilities. From making a productive start to harnessing growth opportunities, each facet of effective management is critical to both individual and team achievements. The managerial journey is filled with opportunities to lead, motivate, and guide your team, as well as build invaluable alliances outside your team.

Success as a manager is not merely about individual excellence but the ability to orchestrate, lead, and inspire the collective brilliance of a team. Follow the guidance in this chapter to elevate your management capabilities. Commit to regular self-assessment, seek feedback, and be daring enough to lead teams to new heights.

Ultimately, the journey toward effective management is a continuous process marked by learning, adaptation, and dedication. Successful completion of this journey can pave the way for you to expand your responsibilities and become an effective leader within your organization. And, incidentally, this is also the topic of the next chapter—effective leadership.

7

Becoming an Effective Leader

In the previous chapter, I discussed everything about becoming an effective manager. I talked about *managing* people, projects, and your own time, among other things. Note the emphasis on the word *managing*. Management is about planning, organizing, and controlling the different parameters and resources to achieve set targets. Leadership, however, goes beyond traditional management tasks. You don't need to formally become a manager to start leading a group of people.

From an organizational perspective, leadership involves mentorship, coaching, and setting a visionary course. It is about motivating and influencing people to work together instead of setting expectations about what they should and should not do. In software engineering, leaders inspire innovation, set the direction, and can emerge at any level of an organization, including the managerial level.

Let's refer back to Chapter 3's 3E's model (enable, empower, and expand) for helping engineering leaders to instill effectiveness in their teams, departments, or organizations from the ground up. As a manager, you can *enable* effectiveness in your team by defining what effectiveness means in your context and facilitating the necessary steps, such as training and measurement, to initialize it. You can also *empower* your team to become effective by organizing the resources and support they need and removing blockers. However, to truly *expand* effectiveness by motivating people to want to become effective, you have to scale yourself up as an effective leader.

In this chapter, we will explore how developing leadership qualities enhances your management skills and contributes to your overall effectiveness. We will discuss a few official leadership roles and avenues generally available in software engineering teams and how they differ in responsibilities. We will also highlight the qualities and traits an effective leader must possess and ways to cultivate these. Finally, we will discuss some fundamental principles to lead and expand your leadership effectively.

Effective Leaders Versus Effective Managers

Before I discuss how to become an effective leader, it is essential to understand how management and leadership are different. It is also important to note that effective leadership does not replace effective management. Rather, leadership and management comprise attributes and activities that are complementary and necessary for success. Figure 7-1 shows how the roles and responsibilities of leaders and managers intersect.

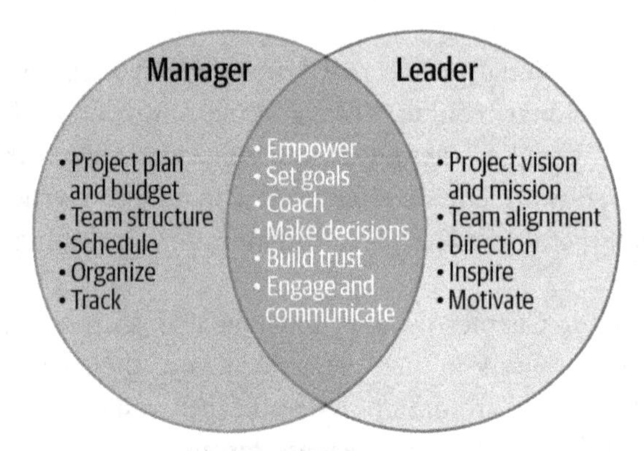

Figure 7-1. Responsibilities of leaders and managers

According to John Kotter (*https://oreil.ly/2zA7C*), a thought leader on the topic of leadership and management, the fundamental distinction between leadership and management is that *leadership produces change and movement, while management produces order and consistency.*

Leadership takes a creative approach to guiding people to success by inspiring and mobilizing teams to embrace new ideas, adapt to evolving circumstances, and reach for innovative solutions. On the other hand, *management* takes a formalized, frameworked approach, focusing on maintaining order and consistency within established structures. It involves planning, organizing, and controlling resources and processes to ensure that day-to-day operations run smoothly, efficiently, and predictably. To understand this difference better, let's look at some focus areas for leaders and managers in Table 7-1.

Table 7-1. Focus areas for effective leaders and managers

Effective leaders	Effective managers
Establish direction: Effective leaders excel in establishing direction. They are visionaries who create an inspiring future vision for their teams. They craft strategies to realize this vision and articulate the big picture.	**Plan and budget:** Effective managers focus on operations and logistics. They specialize in meticulous planning and budgeting. This usually involves establishing agendas, defining timelines, and allocating resources efficiently.
Align people: Effective leaders share their vision with the team so that everyone knows where they are headed. Thus, they provide clarity and purpose that supports cooperation among team members. This inspires teams to work collaboratively toward shared objectives	**Organization:** Effective managers bring structure and order to the operational side of tasks. They excel in organizing and staffing, where they provide the necessary structure for teams by making job assignments and establishing efficient processes. In this way, managers ensure that the work is distributed logically and responsibilities are clear.
Motivate and inspire: Effective leaders are adept at motivating and inspiring teammates, not merely through directives but by empowering them and addressing their unmet needs. They bring a contagious energy that ignites enthusiasm and commitment among their team members.	**Control and problem-solving:** Effective managers are adept at using incentives to motivate and generate solutions to challenges. Managers take corrective actions to ensure smooth operations, preventing and addressing disruptions that might hinder the pursuit of objectives.

This breakdown is intended to highlight the distinctive yet interdependent roles of leaders and managers in achieving organizational success. While leaders set the vision, inspire, and align people, managers provide the essential groundwork for planning, organizing, and controlling to maintain order and consistency. Both leadership and management are vital in their own right, with leadership steering the way for evolution and growth while management provides the necessary stability and structure to support these changes.

Leadership and management need not be mutually exclusive. Effective managers often possess strong leadership qualities. The finest managers seamlessly integrate leadership capabilities to establish a robust operational foundation and elevate their teams through inspiration and vision.

A few approaches to combining managerial responsibilities with leadership qualities are as follows:

Strategic vision
: A visionary approach allows you to align managerial decisions with the broader objectives, ensuring your decisions are directed toward the correct long-term goals.

Motivational leadership
: While managing involves assigning tasks and ensuring they are completed, you can do so effectively by understanding what motivates each team member and assigning tasks and deadlines accordingly.

Empowerment and trust
: As a manager directing tasks, trust your team members to make decisions. This empowerment is crucial for encouraging innovation and creativity.

Adaptability and change management
: While managing involves maintaining order and consistency, a leader must be adept at navigating change. You should be able to guide your team through transitions with a positive attitude toward change.

In essence, the distinction between leadership and management blurs when managers embrace leadership qualities. The best managers recognize that effective leadership is not a separate entity but an integrated approach that enhances their managerial skills, resulting in a more dynamic and thriving work environment.

From this discussion, you can conclude that leadership and management play distinctive yet interdependent roles in achieving organizational success. Leadership propels change and innovation, inspiring teams to embrace new ideas, adapt to evolving circumstances, and reach for innovative solutions. On the other hand, management focuses on order and consistency, ensuring day-to-day operations run efficiently within established frameworks.

I've observed that the most successful managers adeptly incorporate leadership qualities. This dynamic is crucial in software engineering. Organizational hierarchy and job descriptions are tailored to facilitate a collaborative environment where team leaders and managers can complement and support each other.

This ensures clarity for engineers on whom to approach for assistance with specific problems. Leadership roles within software engineering organizations have evolved to foster this collaborative dynamic, a topic I'll delve into in the following section.

Leadership Roles

Organizational structures in software engineering organizations differ widely depending on their culture and priorities. After a person has served as an engineer or senior engineer for a few years and gained the necessary expertise, there are typically two tracks open to them: technical or managerial. Each offers distinct leadership opportunities and requires individuals who can coach and guide their teams through challenges.

In this section, you will look at some typical roles across the industry and what they usually entail in terms of effective leadership. Note that these aren't the only leadership roles in an organization.

Leadership roles in a team depend not only on the overall organizational structure but also on the size and complexity of the project. Larger teams could have one or many technical leads leading the development of different parts of a project. Additionally, such teams would have architects synchronize the efforts led by the technical leads and managers to plan and organize resources. You could also have a product manager who articulates what success looks like for a product and guides the team to make it a reality. Conversely, in small teams, these roles may be combined to have a manager with the technical expertise to lead the team.

Figure 7-2 shows how some of the different types of leadership roles may coexist in a software engineering team.

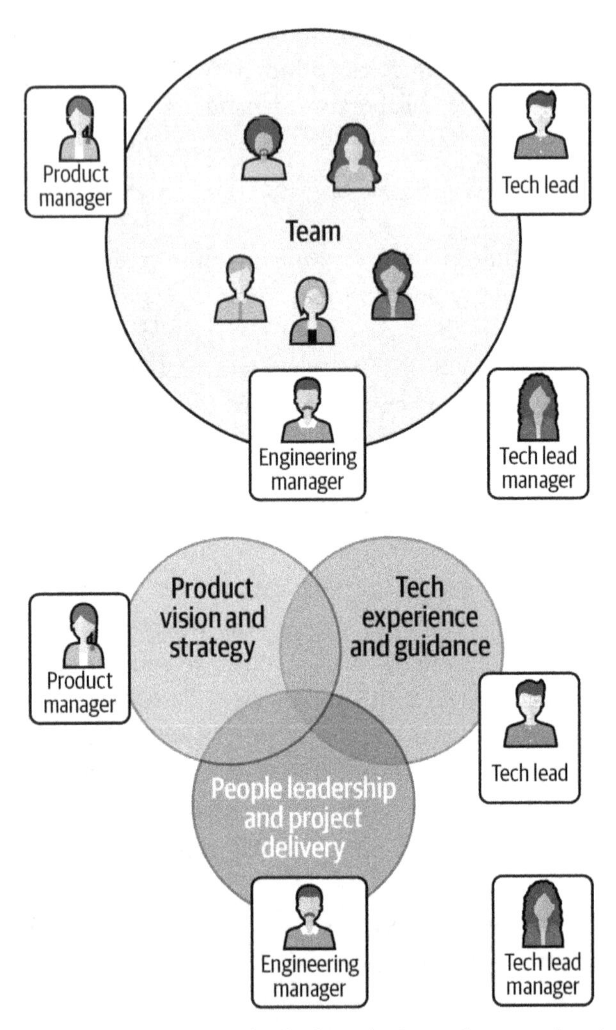

Figure 7-2. Relationships between various leadership roles in a software engineering team

Let's take a closer look at some of these leadership roles.

TECHNICAL LEAD

A *technical lead* is a hands-on role where you provide technical guidance and direction to the engineering team. The designation itself may vary across organizations. It may be a formal title in some workplaces, while it exists more informally in others. In some organizations, the position may be identified as a "software architect," while in others, it may be referred to by titles like "principal engineer" or "lead software engineer."

Irrespective of the name, tech leads play a crucial role in architectural decisions, code reviews, and mentoring junior team members. Technical leads often bridge the gap between the development team and management, ensuring alignment between technical strategies and business goals. Some of the responsibilities of a technical lead include the following:

Guide technical design and architecture
: Tech leads play a vital role in shaping the technical direction of the project by providing guidance on design and architecture. A tech lead must leverage their expertise to ensure that the chosen technical solutions align with the project's goals and adhere to industry best practices.

Set coding standards and best practices
: Tech leads should take the initiative to establish coding standards and best practices within the development team. The tech lead role involves defining and enforcing these guidelines to contribute to overall code quality, maintainability, and consistency.

Lead troubleshooting of complex bugs and issues
: Someone in the tech lead role leads the investigation and resolution of intricate technical issues and bugs. Their deep understanding of the codebase empowers them to troubleshoot effectively, ensuring the stability and reliability of the software.

Make key technical decisions with engineering trade-offs
: Tech leads are responsible for making critical technical decisions, carefully weighing engineering trade-offs to align with project objectives. They consider factors such as performance, scalability, and maintainability to ensure the overall success of the software.

Do hands-on coding alongside the team
: Despite their leadership role, tech leads often find themselves actively engaging in hands-on coding alongside their team members. This approach helps them mentor other engineers while staying connected with the codebase.

Serve as a mentor for development skills

Tech leads also act as overall mentors, guiding team members to enhance their development skills. They lead by example to foster a culture of continuous learning and professional development within the team.

Ensure deliverables meet the quality bar

Tech leads are accountable for the quality of deliverables, ensuring that the software meets established standards and requirements. They conduct thorough reviews and quality assessments to guarantee that the end product aligns with the defined quality bar.

Depending on the size of the project, the scope of these responsibilities will vary—from overseeing a single development team to having cross-team responsibilities.

ENGINEERING MANAGER

An *engineering manager* typically oversees a team of software engineers, ensuring the successful delivery of projects. They are responsible for project planning, resource allocation, team productivity, performance, and career development, including that of the tech lead. This role often involves a mix of managerial tasks, such as performance evaluations and career development, along with technical oversight. In some companies, engineering managers may also be referred to as "development managers" or "technical managers." To recap, an engineering manager's key responsibilities include the following:

People management

Engineering managers should gear up to develop their skills in hiring, talent development, coaching, and mentoring. Engineering managers actively engage in the recruitment process, nurture their team members' potential, provide guidance, and foster a culture of continuous learning within their team.

Manage processes

Engineering managers orchestrate critical processes such as sprint planning, retrospectives, and regular one-on-ones. They should ensure these processes are not just executed but tailored to their team's needs, promoting collaboration, communication, and continuous improvement. They need to check that processes are not sidestepped.

Align team with organizational priorities

Engineering managers must ensure that their team is aligned with the broader organizational priorities. This involves effectively communicating context, goals, and expectations to team members while also shielding them from unnecessary distractions. By serving as a bridge between the team and the larger organization, the engineering manager helps team members focus on their work and deliver value.

Unblock resources

Engineering managers must actively work on unblocking resources needed for execution. They liaise with other departments, manage dependencies, and ensure that their team has the necessary tools, resources, and support to deliver on their commitments.

Technical oversight

While the engineering manager may not have any hands-on coding time, they should maintain their technical acumen. They engage in architecture discussions, ensuring technical decisions align with best practices and organizational goals. This technical oversight helps them guide their team to find sound technical solutions.

Stakeholder interaction

Engineering managers should engage with stakeholders, including having direct interactions with customers. They must understand project requirements, ensure proper communication channels, and act as a conduit between their team and external stakeholders. Engineering managers ensure that the team receives clear requirements from stakeholders.

Strategic work prioritization

Engineering managers must strategically prioritize work aligned with their team and company's vision. This involves balancing project commitments with essential operational work, addressing technical debt, performing and maintenance in line with the organization's strategy.

As you take on an engineering manager role, remember that you must broaden your responsibilities to include comprehensive people management, process leadership, and strategic alignment with organizational goals in addition to technical oversight. Unblocking your programmers is also an essential but slightly underrated aspect of managerial responsibilities.

Joel Spolsky, the cofounder of Stack Overflow and creator of Trello, once said, "Your first priority as the manager of a software team is building the development abstraction layer."[1] He further explains that if a developer is directly exposed to infrastructure issues like access to the project repo on GitHub or overriding a firewall for necessary project work, then the abstraction has failed.

TECH LEAD MANAGER (TLM)

Tech lead managers (TLMs) are rare in many organizations. In Google, small or nascent teams usually have a TLM who can oversee a group of engineers, guiding them in project execution and ensuring the team's productivity. This role involves a mix of technical leadership, project management, and people management. You will need a solid technical background to take up this role and should be able to contribute to technical discussions easily. You should be involved in technical design and communicate relevant design decisions to other teams and stakeholders.

TLMs are responsible for setting priorities, resolving technical challenges, and fostering a collaborative team culture. This role offers the opportunity to do both technical execution and people leadership. But it also comes with the challenge of balancing the two areas while not shortchanging either one. To help with this, TLMs will usually have a smaller number of direct reports as compared to engineering managers. TLM responsibilities include the following:

Blending people management with hands-on technical leadership
: TLMs must balance their responsibilities as people manager and technical leader. This involves not only overseeing the professional development of the team but also actively participating in the technical aspects of projects, setting an example for team members.

Coach and develop engineers on coding skills
: From a people management perspective, part of the TLM's responsibility is nurturing their team, coaching, providing constructive feedback, and guiding engineers to enhance their technical proficiency. TLMs must also ensure individual contributors are challenged in their work and are on track to reach their personal career goals.

---

1 Joel Spolsky, "The Development Abstraction Layer," April 11, 2006, *https://www.joelonsoftware.com/2006/04/11/the-development-abstraction-layer-2*.

Establish technical standards and architecture
TLMs are responsible for setting technical standards and architecture. This entails defining and maintaining coding practices, architectural principles, design, and code reviews.

Help unblock developers when they are stuck
TLMs play a crucial role in unblocking developers when they encounter challenges. This involves providing technical guidance, removing impediments, and keeping upper management appraised of the project's progress and resource needs.

Focus on higher-priority technical work
Sometimes, TLMs may need to concentrate on higher-priority technical initiatives. This could even involve hands-on coding or debugging. TLMs may have to delegate specific people management tasks to balance the other demands of their role. This strategic delegation ensures that both aspects of their role receive adequate attention.

Advocate for the team while coordinating cross-functionally
As the advocate for their team, TLMs engage in cross-functional coordination. This includes representing their team's interests, ensuring effective communication across departments, and fostering collaboration to achieve collective goals.

Make technical decisions weighing various constraints
TLMs are decision makers in technical matters, which involves considering multiple constraints. This includes assessing factors such as project timelines, resource availability, and technical debt to make informed decisions that align with both short-term goals and long-term sustainability.

Provide mentorship and guidance
TLMs play a crucial role in mentoring and guiding team members to enhance their technical skills and professional development. By dedicating time to mentorship, TLMs foster a culture of continuous learning and growth within the team.

As you can tell from the preceding list, having really strong technical aptitude is critical in a TLM role. A TLM often asks intelligent questions and pushes the team to find answers. TLMs communicate a lot with various people, some of whom are purely technical and others of whom are business oriented. TLMs will thus have to switch their communication style constantly. A sign of success as a

TLM is effectively balancing all the responsibilities while finding some extra time to write some code occasionally.

While there may be other roles or other names used to refer to these roles among software organizations, I have tried to discuss the key responsibilities of a team leader or manager in an engineering team in this section. However, responsibilities don't dictate your ability to perform them. How do you know you have what it takes to lead your teams effectively? Find out by assessing yourself on key leadership traits in the next section.

Assessing Your Leadership Skills

Whether you are an aspiring or seasoned leader, some distinctive qualities and characteristics (see Figure 7-3) can tell you where you stand on the effectiveness scale. Beyond titles or roles, these leadership traits encapsulate the essence of what makes someone a capable and inspiring leader. Whether innate or developed, leadership traits shape how individuals navigate challenges, inspire teams, and drive organizational success.

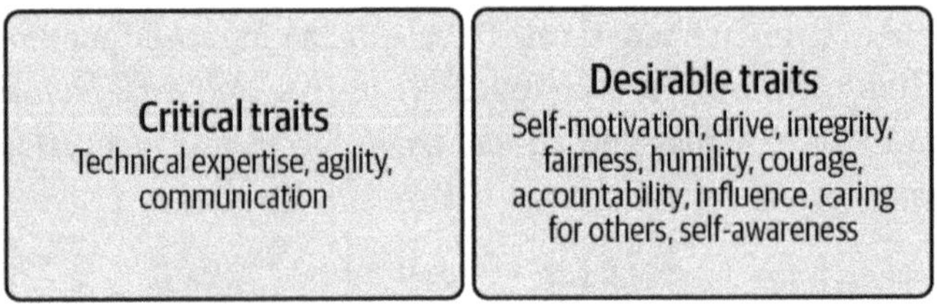

Critical traits	Desirable traits
Technical expertise, agility, communication	Self-motivation, drive, integrity, fairness, humility, courage, accountability, influence, caring for others, self-awareness

Figure 7-3. Assessing your leadership skills

This section explores the key characteristics that distinguish exceptional leaders and offers insights into self-assessment to help you determine where you stand. I will start with the most essential traits critical to tech leaders and then move on to other personality traits that can shape your effectiveness as a leader.

CRITICAL TRAITS

Critical traits are essential qualities or characteristics that enable individuals to excel and make significant contributions. In the context of engineering leadership, critical traits are the defining attributes that distinguish effective leaders from those who struggle to inspire and guide others. In tech leadership, a few traits are particularly crucial. Technical expertise and problem-solving capabilities allow leaders to make informed decisions and provide valuable guidance on complex projects. Agility is vital, as leaders often need to guide teams through technology updates and shifting project scopes. Clear communication is essential for bridging the gap between technical teams and stakeholders who may not share the same technical background, ensuring everyone is aligned on objectives and expectations.

Technical expertise

Technical expertise is a critical skill for any software engineer, but it is especially important for team leaders. As a team leader, you will mostly be using your experience and knowledge to help others in your team to solve the problems they encounter. You are responsible for helping your team identify, analyze, and solve problems within the constraints of the project.

Complex problems often require the technical expertise to come up with creative and innovative solutions that are also feasible. You should not only be able to think out of the box, but with your past experience, you should be able to quickly analyze which of the solutions will actually work in the long run. When the timeline is tight, questions are thrown at you by different team members simultaneously, and you have to efficiently juggle them without letting the ball drop. Some questions you can ask yourself to assess your technical expertise are as follows:

Do I have the required technical know-how for this project?

Since technology is always changing, it is likely that team leaders may never have worked with some aspects of the technology being used on the project. As long as your foundations are solid and you have worked on something similar, you should be good to go. However, if you are someone from a web app development background leading a mobile app project for the first time, it could be a problem.

Have I perceived a team member's face light up when I guided them toward a creative solution?

It's magical when you guide your team members to uncover a solution that they could not have imagined by themselves. It's as if a light bulb is switched on, and they look at you with a lot of respect.

When was the last time I helped someone solve a difficult problem?

You hone your technical skills when you frequently get your hands dirty solving technical problems. Effective leaders love attacking problems with passion. If you are one of them, your answer to the preceding question will be no more than three months.

How do I approach challenges? Am I open to trying unconventional ideas?
: Problem-solving, especially with complex problems, requires thinking beyond the normal and exploring different ideas. Interesting problems demand interesting solutions, and you have to let yourself find them and try them.

Agility

Agility refers to your ability to learn, unlearn, and adapt yourself to changing conditions. Agility helps you quickly revise and review your strategies in times of uncertainty and come up with a way forward. You must be able to think strategically, learn quickly, and change your course to convert problems into opportunities.

Situational awareness is essential for agility, as it enables you to be proactive instead of reactive. Leading with agility improves your ability to respond to events quickly while making the right decisions. One of the key requirements for becoming agile is not being afraid to question your own beliefs and assumptions but instead being open to trying new things and making mistakes. Consider asking yourself these questions to assess your agility:

How rapidly do I acquire, apply, and share knowledge?
: The faster you are able to assimilate and forge newly acquired knowledge with your previous experience and foundations, the greater your agility will be. This could be a new business paradigm, technical breakthrough, or productivity tool.

How do I approach new information?
: The best way to approach new information is with curiosity and a willingness to understand. It may be something relevant to your work or not, but there is only one way to confirm this.

How quickly am I able to take action based on changing situations?
: Change could present itself in different forms, such as a change in project requirements or a sudden resignation in the team. You should be able to quickly adapt to these situations and take corrective action.

Since the COVID pandemic, there have been many changes in how we work. This has been a significant test in terms of leadership agility. Leaders had to first adapt to the switch to remote work and then gradually to a more hybrid work environment. An excellent example for leadership adaptability in this regard came up during a discussion with an ex-colleague.

When the pandemic hit in early 2020, Sandra, an engineering director at a midsize multinational corporation, knew she needed to act quickly. With over 50 engineers across Asia, Europe, and South America, shifting to fully remote work seemed daunting. However, Sandra approached this challenge as an opportunity to reimagine team connectivity.

She started by having one-on-one conversations to understand everyone's unique situations and concerns with remote work. Sandra then ran an online poll on policies for flexible schedules, meeting times, and communication norms—empowering the team to shape this transition. Throughout this process, she had been open to suggestions and change.

When productivity dipped months later, Sandra responded adaptively. She listened to understand factors impacting engagement and recalibrated based on insights around work-life balance, technical roadblocks, and pandemic fatigue. She evolved her approach, providing more autonomy while guiding the team—not micromanaging it. As a result, her team continued to thrive remotely even after the pandemic had ended, demonstrating the power of adaptive leadership.

Thus, we see that agility comes not just from following a prescribed methodology but rather from being able to navigate any situation strategically with minimal loss of momentum.

Communication

All the other skills mentioned here are only relevant if you are able to communicate with your team. Clear communication is essential for you to share your vision and help others understand it. Communication, both verbal and nonverbal, enables you to demonstrate other qualities like integrity, humility, and technical expertise. It is the most essential tool in your leadership toolbox that helps you motivate and influence others. To assess whether you are a good communicator, analyze your listening, explaining, persuading, and contextualizing skills and ask yourself the following questions:

Does my team always understand me?

Communication should be both clear and concise for it to be understood by others. Beating around the bush can confuse others. If you are a good communicator, your team will understand you most of the time.

Have there been any instances of misunderstandings or miscommunication where a team member did not understand what I was trying to convey?
: If there have been such instances, try to find out what went wrong and how you could have improved the communication.

Do I tailor my communication style to the individual or group I am addressing?
: The same message can be conveyed in many ways depending on who the audience is. Junior or senior engineers, business or technical personnel, stakeholders, and management—communication for each of these groups needs to be tailored to their background and what is most relevant to them.

Do I often need to repeat instructions?
: This can be a problem if you talk very fast, are not loud enough, or have a heavy accent. Again, identify whether there is an issue and try to correct it through training and practice.

Do I actively listen and remember the discussion in the right context?
: Listening is an important part of conversation. Allow others to speak, show an interest in what they are saying, and try to remember the gist of what was said for future reference.

Despite what your assessment tells you, *there is always room for improving your communication*. Even when you *think* you have a plan or strategy for good communication, there are almost always opportunities to improve.

Case in point: I remember a significant launch for a developer-focused product that my engineering team and I were working on at Google. We had worked with product, marketing, business development, public relations, our management, and leadership to try to keep everyone looped in. We crafted what we thought was a clear, concise message and worked hard to get sign-offs from everyone ahead of the "moment." There was just one problem: despite being developer focused, we had completely neglected to inform our Developer Relations (DevRel) team about the launch.

At the time, we did not have a dedicated DevRel team member assigned to the area we were working on, but we absolutely could have given them a heads-up about it just in case they had feedback. Because it came as a surprise, they had to scramble at the last minute to provide input, which we felt awful about. The launch went well in the end, but we learned to take an oversharing/inclusive approach to internal launch preparations in the future to avoid this.

One of the well-known tools for effective communication is the 7 C's: your communication should be clear, concise, concrete, correct, coherent, complete, and courteous. I've always liked this tool and tried to remind the leaders on my team to use it. The 7 C's are a good starting point for clear meetings, emails, conference calls, reports, and presentations so your audience gets your message.

In this case, I could have been more complete in the communication, and working with DevRel from the beginning would have helped us refine our messaging without that last-minute scramble.

Improve your leadership skills

To improve your leadership skills, consider the following tips:

Agility
: Regularly engage in learning opportunities, such as workshops, conferences, or online courses, to stay updated with the latest industry trends and technologies. Embrace change and be open to adapting your strategies when needed. Cultivate a growth mindset, viewing challenges as opportunities for learning and development. Encourage your team to do the same, fostering a culture of continuous improvement and innovation.

Communication
: Practice active listening, paying close attention to what others are saying and asking clarifying questions to ensure understanding. Seek feedback from your team and stakeholders regularly, and be open to constructive criticism. Tailor your communication style to your audience, considering their preferences, backgrounds, and levels of understanding. Regularly engage in open and transparent communication with your team and stakeholders, keeping them informed of project progress, challenges, and successes.

Empathy
: Put yourself in your team members' shoes, striving to understand their perspectives, feelings, and needs. Show genuine concern for their well-being, and offer support and resources when needed. Foster an inclusive environment where everyone feels valued, respected, and heard. Lead with compassion, recognizing that each individual has unique strengths, challenges, and aspirations.

Vision

Develop a clear and compelling vision for your team, aligned with the organization's goals and values. Communicate this vision effectively, inspiring your team to work toward a common purpose. Regularly review and refine your vision, adapting to changing circumstances and new information. Encourage your team to contribute to the vision, fostering a sense of ownership and commitment.

Delegation

Empower your team members by delegating tasks and responsibilities appropriately. Provide clear expectations, guidelines, and resources, while giving them the autonomy to approach tasks in their own way. Trust in their abilities, and offer support and guidance when needed. Regularly review and adjust your delegation approach, ensuring that team members are challenged and growing in their roles.

Integrity

Lead by example, demonstrating honesty, transparency, and ethical behavior in all your actions. Keep your promises and admit when you make mistakes. Hold yourself and your team accountable to high standards of integrity, fostering a culture of trust and respect. Make difficult decisions with courage and conviction, always keeping the best interests of your team and the organization in mind.

By focusing on these key areas and consistently working to improve your skills, you can become a more effective and impactful leader.

DESIRABLE LEADERSHIP TRAITS

In addition to these essential traits that contribute to a leader's success, there are others that can multiply the impact of your leadership. These qualities can shape your character and influence how you interact with your team, make decisions, and navigate challenges. Let's now discuss these integral traits in detail, examining their significance in fostering a positive and dynamic leadership approach.

Self-motivation

Self-motivation is the internal drive that pushes you to take action toward your goals. It's the ability to find the energy and willpower to do something, even when you don't feel like it. It's what keeps you going when you face challenges or setbacks. When you are self-motivated, you do more than check things off a to-do list. You know what you are really after, so your sense of motivation comes from within.

To be an effective leader, you must regularly assess your motivation to lead the team. Reflect on your goals for the team and whether you have a consistent inner drive to achieve them. Ask yourself these questions:

Am I motivated to lead this team?
Do you look forward to leading the team and getting things done when you start your day? If you prefer to avoid it, then probably you are not motivated. Find out what is missing and act on it.

Do I believe in our goals and our abilities to achieve them?
Do you think your goals are tangible and achievable, or is your mission statement just that—a statement? Belief is essential because it gives you the confidence and motivation to work toward goals.

Am I passionate about the vision we are working toward?
Passion propels you to work hard with dedication to achieve your vision and is a crucial component of self-motivation.

Drive

Slightly different from self-motivation, *drive* is the ability to maintain focus on the end goals, avoid distractions, and, in Dory's famous words from *Finding Nemo*, "Just keep swimming." It is the unwavering determination and passion to pursue goals, achieve objectives, and overcome obstacles. It is the motivational force that propels you to make a difference. It is characterized by relentless energy, a solid belief in your vision, and a strong commitment to excellence.

Driven leadership is not about blind ambition or reckless pursuits. It is about harnessing your passion, determination, and focus to make a positive impact. Questions that can help you assess your drive are as follows:

Am I distracted easily from my end goals?
If you are driven and focused, then it is difficult to distract you while you pursue your goals. If you readily agree to participate in meetings or activities that do not contribute to your commitments, then you probably lack drive.

Am I able to follow up relentlessly and course-correct quickly?
While a laid-back or casual leadership style lets you connect with the team members in a stress-free space, allowing them to evolve independently, you

must follow up relentlessly if you feel they are falling behind. Relentless follow-up with external parties may also be required to unblock your team if needed.

Do I remain tenacious and optimistic in the face of challenges?

Survival can be difficult if challenges and setbacks affect your enthusiasm for leading the team. Instead of getting disheartened, learning from the mistakes, picking up the pieces, starting afresh, and motivating your team to look forward are essential.

Am I able to recognize and seize opportunities that can help me with achieving my goals?

If you are driven and constantly aware of your goals, it becomes easy to spot opportunities that can help your team. For example, a piece of technology being developed by another team that can be reused by yours is an opportunity to reduce effort and time.

Integrity

Integrity is the quality of being honest, truthful, and ethical in one's actions and decisions. It is the foundation of trust and respect and is essential for effective leadership. It's a highly valued trait, since your employees look up to you as a role model. By demonstrating integrity through your actions and words, you can set the tone for your team.

While deciding to act with integrity is excellent, how do you know you are really following through on that promise? For example, you may have said that you believe in transparency, but do you demonstrate it? One way to assess yourself is to examine past actions and decisions to identify congruence with stated values. A few questions that might help are these:

Have I consistently acted in alignment with my stated values?

If you preach a particular value to your team, you should also practice it. For example, if you expect the team to be punctual or work from the office at least three days a week, do the same yourself.

Did I ever fail to deliver on my promises?

If you have promised someone on your team a particular opportunity during a one-on-one and you fail to deliver on that promise, it will impact the overall perception of your integrity. If this happens repeatedly, it will impact the team's morale.

How transparent am I with others?

There is a certain maturity in transparency and being up-front in the way you deal with others. When you are transparent with information, you are letting your team know that you trust them to take things in the right spirit.

Fairness

Fairness is related to integrity. A fair leader ensures that all individuals in the group are treated fairly and impartially. This includes distributing resources and opportunities equally and ensuring everyone has a voice. You are fair if you base your decisions on merit, not personal biases or preferences. It is important to create a culture of trust and respect where employees know their efforts are being noticed. Fairness also encourages healthy competition in the team. Ask yourself these questions:

Do I always make objective decisions based on facts?

There may be situations where two or more team members propose different solutions to a problem. If you make fact-based decisions, you will assess the merit of each proposed solution based on the nature of the solution rather than on who is proposing it.

Do I hold any prejudices that may subconsciously influence my decisions?

Team members may come from different educational, financial, or personal backgrounds. Some of these backgrounds may appeal to you because of some preconceived theories. Fairness implies that you don't bring these theories to the table when making decisions involving these team members.

Am I impartial when delegating responsibilities?

You may have favorites when it comes to delegation because a couple of team members perform the delegated tasks exceptionally well. But everyone deserves a fair chance, and you are the only one with the power to give them that chance.

Humility

Humility is the ability to recognize your limitations, acknowledge your mistakes, and be open to feedback. It is the opposite of arrogance and self-importance. Humble leaders are not afraid to admit when they are wrong, and they are always willing to learn from others. You are more likely to be trusted by your employees if they realize you are humble, genuine, and approachable.

Humility enables you to learn and grow continuously and to receive and accept feedback naturally. It is also a great trait to have when collaborating with others. Are you humble? Ask yourself these questions:

Am I comfortable admitting when I make mistakes or don't have all the answers?
: You can't know everything. Rather than pretending to be a know-it-all and risk giving the wrong answers, try to find the right ones with your team.

Am I open to learning from others in my team, including juniors?
: Learning can come from anywhere. In a diverse team, everyone has something to offer. Eagerness to learn and willingness to listen will help you accumulate knowledge and grow, and they will also make you more approachable.

Do I give credit where it is due?
: You couldn't have accomplished a win all by yourself if you had a team backing you up. If you hog credit instead of sharing it with your team, you are certainly not humble.

Courage

As a leader, you are often required to make difficult decisions and stand by them. This is only possible with a good dose of courage. Every innovation is born out of the courage to step out of the comfort zone and meet a challenge head-on. Courageous leaders are not afraid to take risks or start from scratch if the situation demands.

As per Aristotelian ethics (*https://oreil.ly/KCQ9i*), courage is a virtue between cowardice and rashness. This notion of courage remains relevant today, as it highlights the importance of facing challenges and making difficult decisions with a rational and balanced approach. While courage may sometimes be interpreted as the lack of fear, it also implies being wise and knowing when and for what reason you should be fearless. In simple words, take risks but calculate the odds beforehand. To assess if you are courageous as a leader, ask yourself these questions:

How comfortable am I with uncertainty?
: There is always an element of uncertainty when executing projects. Requirements may change, or the business may be affected by regulations. Courageous leaders embrace this uncertainty and are prepared for the future.

Have I challenged existing norms?

Even if your team has been doing things a particular way for a considerably long time, there is always scope for optimizing and streamlining processes. You should have the courage to implement new strategies and optimization techniques that you think will be helpful.

Am I willing to take smart risks and question the status quo?

There is always risk involved when you change something that works. You need to analyze this risk and compare it to the potential returns. If you are sure that the returns outweigh the risks, then you must be willing to take those risks.

Accountability

Accountability is another trait essential for building trust and fostering transparency in a team. *Accountable* leaders own their decisions, whatever the outcome. To become accountable, you must take responsibility for the results of your work.

Remember that you are responsible for the people who report to you, too. So if a project misses its deadline because a key person on your team is not well, you are still ultimately accountable because it is your responsibility to have a contingency plan in place. At the same time, in times of success, you must acknowledge it as a team effort rather than taking credit for the win yourself. A few questions you can ask yourself to find if you act with accountability are as follows:

How readily do I own my mistakes?

If one of your decisions leads to a setback, you must accept responsibility for that setback. By doing so, you set an example for the team to own their mistakes.

Do I take responsibility for my decisions and actions?

You must stand by your decisions and actions without offering any excuses when questioned by higher management or stakeholders. Do not pass the buck to your team.

Do I even mildly indulge in finger-pointing or passing the blame?

When something goes wrong, the last thing you should do is point fingers. Accept what happened with sincerity, correct the course, provide feedback where necessary, and move on.

Influence

Your influence on others is what makes you a true leader. You can say that this influence is the result of your other leadership traits. People will follow you because they want to be like you. You can change behaviors and inspire others to take action. Because someone sees you as a role model, you can affect their attitude, beliefs, behavior, and development.

Influence is not about manipulation or control. Instead, you should be able to motivate and inspire your team members to see the value of a shared goal and work together to achieve it. Influential leaders are able to tap into the strengths and motivations of their team members and create a positive and productive work environment. Find out if you are influential by asking yourself these questions:

Do people listen to me because of my authority as a leader or because they trust me to do the right thing?
: If you are influential, people will want to follow you not because of your title or role but because they respect you and want to emulate you.

Would I be able to guide others even without authority?
: You should not need any authority to give the right answers. What you need is confidence in your knowledge and experience.

Caring for others

Effective leaders demonstrate a genuine concern for the well-being of their employees and go out of their way to support them. They exhibit empathy, compassion, and understanding, creating an environment where everyone feels valued, supported, appreciated, and heard. By cultivating these qualities, leaders can build stronger relationships with their team members and foster a positive and inclusive work culture.

Do you think you are an effective leader? Find out by asking yourself these questions:

Am I aware of what's going on in my team members' lives?
: Are you able to talk with your team members generally about things other than work? This does not have to be personal information. It could be the latest movie they saw or a book that they liked.

Am I able to empathize with others?

You should be able to relate to specific situations that your employees find themselves in. For example, if you have someone in your team who has recently emigrated from another country, you should understand that they may need time to settle in both personally and professionally. They may need help understanding specific local terms or may experience difficulties in getting to work on time.

Are my team members overly cautious while communicating with me?

If your team members think you are unconcerned about their well-being, they may simply avoid discussing details with you. They may perceive you as a superior, someone not to be mingled with.

Self-awareness

Finally, if you have honestly and accurately been able to assess where you stand with respect to all of the aforementioned traits, then you are self-aware. *Self-awareness* is the ability to understand oneself, including one's strengths, weaknesses, motivations, and values. It takes time and effort to become more self-aware, but the benefits are well worth it. Self-awareness is essential if you wish to be effective.

It would not be right to ask you to assess your self-awareness by asking questions. A better idea is to seek feedback from others who you know will answer you with honesty. Thus, the only question you need to ask yourself to assess your self-awareness is: does my assessment of my strengths and weaknesses match with theirs? If there is a close enough match, then you are self-aware; if not, then you have a lot of work to do.

Leading Effectively

You have now explored the various leadership avenues and precisely what they mean. You have also learned about the traits essential for effective leadership. However, it's important to understand that the core principles of effective leadership in software engineering transcend job titles and specific roles. I am referring to a collection of practices that can elevate teams and propel organizational success. It's a dynamic blend of skills and strategies that, when mastered, fosters an environment where teams thrive. Whether you're a seasoned manager, a technical lead, or an aspiring leader, the essence of effective leadership lies in the ability to navigate challenges, inspire innovation, and cultivate a culture of continuous improvement. Understanding and embodying these core leadership principles and practices is paramount for steering teams toward excellence.

LEADERSHIP STYLE

Different leadership styles can be employed to inspire and motivate teams toward achieving common goals. Transformational, democratic, and servant leadership styles are three of the most prominent approaches to leading teams and organizations. Each style emphasizes different qualities and behaviors, resulting in varying impacts on team dynamics, decision-making processes, and overall organizational success.

Transformational leadership is characterized by a leader's ability to inspire and motivate their team to achieve extraordinary results. Transformational leaders are visionary changemakers who inspire their teams to pursue ambitious goals and achieve remarkable results. They possess the ability to articulate a compelling vision, motivate individuals to transcend their self-perceived limitations, and instill a sense of collective purpose. Transformational leaders challenge the status quo, encourage risk-taking, and foster a culture of innovation.

Democratic leadership, also known as participative leadership, emphasizes involving team members in the decision-making process. Democratic leaders foster a collaborative environment where team members feel valued and respected. By sharing information and seeking feedback, democratic leaders create a sense of transparency and accountability. This collaborative approach leads to more informed decisions and increased buy-in from team members.

Servant leadership prioritizes the needs and well-being of team members, focusing on empowering them and developing their abilities. Servant leaders focus on creating an environment that supports individual growth and development. By empowering their teams and fostering a culture of service, servant leaders inspire their followers to achieve their full potential. This focus on the team's well-being leads to increased engagement, motivation, and productivity.

Choosing a leadership style

The choice of leadership style depends on various factors, including the leader's personality, the team's dynamics, and the organizational culture. Effective leaders often employ a combination of these styles, adapting their approach to suit the specific situation and context.

Personally, my style is closest to *servant leadership*. As a leader, my aim is to empower and support individuals in reaching their full potential. The key aspects of servant leadership according to me are as follows:

Empathy
: You must be deeply attuned to the needs and concerns of your team members. You prioritize understanding the perspectives of others to provide meaningful support.

Humility
: You acknowledge your limitations and value the contributions of each team member. You create an environment where everyone's expertise is recognized and appreciated.

Stewardship
: You view yourself as a steward of the team's well-being and are committed to fostering a positive and growth-oriented culture.

Commitment to development
: You actively invest in the development of your team members. This involves providing mentorship, guidance, and opportunities for skill enhancement.

Servant leadership is a powerful model that aligns with the idea that effective leaders prioritize the success and growth of those they lead. It's about creating a culture of service, where leaders focus on enabling others to thrive and contribute their best to the collective goals of the team or organization.

To understand how servant leadership works practically, consider Jake, a leader working closely with his team during sprint planning. While formulating a plan, Jake guides the team in breaking down the items from the project's backlog into smaller, manageable tasks. He encourages team members to estimate the effort required for each task, ensuring that the workload is balanced and achievable within the sprint's timeframe. Simultaneously, he helps the team members identify and address any potential roadblocks that could hinder their progress. He consults with team members to find solutions and ensures that they have the necessary resources and support to accomplish the sprint's goal.

Combining different styles

You can also mix leadership styles to adapt to different situations. This is known as *situational leadership*. The idea is that the most effective leadership style depends on the specific situation, context, and needs of the team. For example, consider a software development project where the team is working on a complex and innovative feature. The leader might adopt a transformational style to inspire creativity and set a compelling vision. As the project progresses and the team encounters challenges, the leader might switch to a servant leadership style to provide the necessary support and remove obstacles.

I remember a time when I was leading a team at Google working on performance features for Chrome, and we had to put this principle into practice. We were in the thick of an ambitious project, and I noticed something that I've seen time and again in software engineering management: the struggle to scale oneself as a leader. Here's how it unfolded.

We had a team member, let's call him Mark. Mark was brilliant, a real star in coding and problem-solving. But as he transitioned into a managerial role, he faced a common pitfall: he tried to do it all. He was coding, managing, reviewing, and planning. It was a classic case of a manager not knowing how best to scale themselves.

The turning point came during a particularly intense sprint. Deadlines were looming, and the team was feeling the heat. Mark was stretched thin, trying to maintain his usual level of involvement in coding while managing his growing team responsibilities. It wasn't sustainable, and it showed in both the team's morale and output.

That's when I stepped in, bringing in some lessons from my own journey. Here's what we did:

Delegation is key
: We identified Mark's unique strengths and the tasks that only he could do. Everything else was delegated. It was important for him to trust his team members with these tasks, even if it meant they might do things differently than he would.

Ruthless prioritization
: Together, we went through his responsibilities and trimmed the fat. We asked, "Is this task moving us toward our main goal?" If not, it was either dropped or postponed. This helped Mark focus on what was truly essential for the team's success.

Embracing a coaching mindset
: Instead of diving into coding himself, Mark began mentoring his team members, empowering them to tackle complex problems. This not only enhanced the team's overall skill set but also freed up his time for strategic thinking and leadership tasks.

Setting clear boundaries

Mark learned to set boundaries around his time. This meant dedicated slots for deep work, team meetings, and his own learning and development. A balanced schedule improved his effectiveness and set a good example for his team.

Regular reflection and adaptation

We established a routine for Mark to reflect on his management style and its impact. This ongoing process helped him adapt and grow as a leader, fine-tuning his approach to the needs of the team and the project.

Some of the changes I mention here are leadership principles that we will discuss in detail in the following sections. For now, let's look at the result of these changes. We found that not only did Mark become a more effective manager, but the team's productivity and satisfaction also soared. Team members were more autonomous, engaged, and aligned with the project's goals.

In software engineering, especially in high-stakes environments like Google, this combination of delegation, prioritization, and adaptive leadership is crucial. It's not just about managing tasks; it's about empowering people and steering the ship in the right direction.

As leaders, especially in tech, we often have to unlearn the notion of doing it all. The real art lies in enabling others to excel, creating a space where the whole team can contribute their best. That's the essence of scaling oneself as a leader.

Environment-based leadership

An important factor that can affect your leadership style is the kind of environment you are working in, mainly the size, scope, and complexity of the organization, product, or project and the urgency with which software needs to be developed. A democratic approach may be suitable for smaller or less complex projects with flexible timelines. Such projects are good learning opportunities for everyone. However, leaders might need to be a little ruthless, taking risks and leading with intent, on complex projects with tighter timelines. Consider the following two fictionalized examples to understand how the project environment can affect leadership.

Startup leadership: Empowering innovation. Meet Nisha, chief technology officer at CodeCrafters, an edtech startup developing virtual reality–based coding tutorials for kids. With a lean team of developers, Nisha recognized their competitive advantage was rapid innovation fueled by talent.

Given scarce resources, Nisha focused ruthlessly on the riskiest yet most pivotal part of the product—the VR simulation engine. She codesigned prototypes with her developers, fully hands-on, granting autonomy with guardrails on coding standards. Nisha embodied entrepreneurial leadership, embracing uncertainty and creating psychological safety so developers felt comfortable trying bleeding-edge ideas.

When the product validation phase exposed integration issues between simulation and assessment modules, Nisha adopted an agile leadership mindset—gathering insights through user tests and then outlining new technical objectives. She aligned the team to improvise and iterate, rapidly based on real-world feedback.

Within months, CodeCrafters launched a novel product that captured a significant market share. Nisha demonstrated how in startups, tech leadership must revolve around facilitating innovation velocity, being adaptable, and bringing out the best in scarce but world-class developers.

Enterprise leadership: Bridging business and technology. Legacy Enterprises, a retail company, wanted to leverage AI-based price optimization to maximize profitability. With a project targeting 50,000 products, Director Malcolm needed to bridge skeptical business units expecting immediate ROI with patient technologists focused on building long-term capabilities.

Malcolm embraced collaborative leadership, coordinating extensively between data scientists and category managers who had conflicting priorities and vocabularies. He role-modeled empathy, seeking to understand constraints on both sides—unrealistic deadlines versus infrastructure limitations. Malcolm carved out dedicated collaboration time for working through interdependencies.

When quick wins were not materializing as expected, Malcolm adapted by introducing process improvements. He set up central coordination committees that ensured clear requirements flow between business needs and technical complexities. Malcolm also nurtured psychological safety by establishing a mutual understanding that this cutting-edge project required patience.

Over 12 months, early AI models were refined based on user feedback into profitable insights. Despite initial struggles, Malcolm's vision and ability to synthesize business impact with engineering excellence proved essential for enterprise growth.

Nisha's and Malcolm's stories show us that while knowledge of different leadership styles is essential, you will have to practice using different styles based on the project situation. With time and experience, you will become skilled at reading what the project needs at different points in its lifecycle and be able to adapt to changes promptly.

As you become more experienced, you will be able to use your knowledge of different leadership styles to choose the most effective style for the situation. This will help you to lead your team to success, even in the most challenging circumstances.

STRATEGIZING

Strategizing is crucial for effective leadership, providing a clear roadmap that aligns the team's efforts with organizational goals. Without a clear strategy, you risk becoming reactive to external forces and losing sight of your purpose while leading a team. Strategizing allows you to proactively shape your team's future by understanding its internal strengths and weaknesses and identifying opportunities for growth and innovation. By developing and implementing a well-defined strategy, you get to effectively allocate resources, align team efforts, and make informed decisions.

Strategizing also fosters a culture of shared purpose and understanding within the team, empowering employees to contribute meaningfully to the team's objectives. Effective leaders recognize the importance of continuous strategic adaptation as market conditions and stakeholder expectations evolve over time. A well-crafted strategy will help you navigate challenges and seize opportunities. Effective strategizing is not just an advantage but a necessity if you want to provide direction and resilience to your team. Here are some things you can do to strategize effectively.

Visualizing the future

Visualization is a powerful tool that allows you to anticipate challenges and make informed decisions. Thus, it aids in risk mitigation, promoting adaptability and innovation. Great engineering leaders envision the future landscape, aligning it with their mission. This vision allows you to stretch your imagination toward what is attainable. For example, foreseeing how emerging technologies align with evolving customer needs and market dynamics helps you plan for training and skill development for the team. Some pointers to 360° visualization are as follows:

Environmental scanning
: Stay attuned to industry trends, technological advancements, and market shifts. Understanding the broader context is crucial for informed strategic planning. Anticipate how emerging technologies might impact your team's work and product landscape. Being proactive in adopting or adapting to new technologies is critical for staying ahead.

Scenario planning
: Develop multiple scenarios that depict different potential futures. Consider various factors that can impact your team. By envisioning different scenarios, you prepare your team to adapt to diverse circumstances.

Risk assessment
: Systematically assess potential risks that could impact your projects or initiatives. Categorize risks based on their probability and potential impact. This risk assessment helps in developing mitigation plans and building resilience into your strategies.

Diverse perspectives
: Seek input from team members with diverse backgrounds and expertise. Having a variety of perspectives enriches your understanding of potential future scenarios and ensures a more comprehensive approach to visualization and strategic planning.

By combining these approaches, you create a comprehensive strategy that not only envisions positive outcomes but also considers and prepares for potential challenges and disruptions.

Defining a strategic roadmap

Visualization helps you create a strategic roadmap. Defining a strategic roadmap involves outlining a clear and cohesive plan, complete with initiatives to be launched and milestones in achieving them. These planned initiatives become your rallying points, while measurable milestones help you gauge your team's progress. For example, you may have identified generative AI adoption for higher development productivity as a focus area for your team in the coming year. Define how this aligns with the company's vision and the initiatives and milestones planned to move forward in this direction. Here are five key dos and don'ts for creating an effective strategic roadmap:

Do maintain clarity and simplicity
: Keep the roadmap clear and straightforward. Use visual aids that are easily understandable by team members and stakeholders. Clarity fosters better communication and understanding.

Do set measurable milestones
: Break down the roadmap into measurable milestones. Each milestone should be specific, measurable, achievable, relevant, and time-bound (SMART). This allows for easy tracking of progress.

Do build flexibility and adaptability into the roadmap
: Acknowledge that priorities might shift and the plan may need adjustments based on feedback, changing market conditions, or unforeseen challenges. For example, in the case of generative AI adoption, technology with higher accuracy may become available, or there could be legal concerns about code safety issues in the future.

Don't treat the roadmap as a static document
: A strategic roadmap should evolve over time based on insights, feedback, and changing circumstances. Regularly update and refine the plan.

Don't lack stakeholder involvement
: Develop the roadmap in collaboration with others. Involve key stakeholders, including team members, managers, and relevant departments, in the creation process. Their input ensures a more comprehensive and accurate plan.

By adhering to these dos and don'ts, leaders can ensure that the roadmap is a valuable tool for communication, alignment, and successful execution of strategic initiatives.

Immersive strategic thinking

Immersive strategic thinking is a dynamic and deliberate cognitive process essential for leaders. Unlike routine decision making, this approach involves dedicated and uninterrupted periods when leaders engage in deep reflection, explore innovative possibilities, and analyze data to formulate transformative initiatives.

As a leader, you must make a commitment to yourself to step away from day-to-day operational demands and immerse yourself in a reflective space. You must create a space and time to conduct extended focus sessions free from distractions. By doing this, you take a proactive stance to solve problems, explore options, and come up with creative solutions. Here are some ways you can facilitate this process:

Strategic retreats
: Organize offsite retreats or workshops focused on strategic thinking. These events provide a change of environment and allow for undisturbed concentration on long-term goals.

Digital detox
: During dedicated strategic thinking sessions, consider a digital detox. Turning off notifications and temporarily disconnecting from the constant influx of information fosters a focused thinking environment.

Quiet spaces
: Designate or create physical spaces within the workplace that encourage quiet and reflection. These spaces can serve as havens for deep thinking away from the usual operational hustle and bustle.

These are some of the ways you can carve out the necessary space and time for immersive strategic thinking. Choose what is suitable for your work environment and create a bubble that helps you bring out the most focused version of yourself.

Ruthless prioritization

What are the most crucial tasks that will take you closer to your desired outcomes? This is perhaps the single most important question that you should answer before assigning priorities. *Ruthless prioritization* is the disciplined and unwavering focus on identifying and concentrating efforts on the most crucial tasks or initiatives while deliberately deprioritizing less impactful activities. It involves making tough decisions about what to pursue and what to defer.

Effective leaders hone in on three to five strategic bets that align most closely with their overarching vision. Each chosen initiative is carefully evaluated for its material contribution to the long-term goals. This deliberate curation serves as a protective shield for the team's bandwidth, ensuring that energy and resources are concentrated where they can have the most significant impact.

The courage to say "no" is a cornerstone of this approach. It is not merely a rejection of tasks; rather, it's a strategic imperative. By declining efforts that don't align with the chosen strategic bets, you can maintain a clear and unwavering focus on what truly matters. This clarity enables teams to pour their efforts into projects that move the needle, fostering a culture of concentrated excellence. It embodies the leadership principle that strategic clarity is the bedrock upon which impactful outcomes are built.

PLAYING THE PART

Leadership is an ongoing process that requires consistent effort and dedication. You cannot decide that you will spend 75% of your working hours leading purposefully and wing it the rest of the time. Effective leaders go all in and must constantly practice their skills and adapt their approach to meet the ever-changing needs of their teams and the industry as a whole. You may have to be intentional about it at first, but over time, the ability to continuously engage as a leader will become ingrained in your psyche.

So, what are a few things that you can intentionally do to continuously play your part as a leader?

Relentless communication

Communication is the cornerstone of effectiveness, and as a team leader, you must tirelessly communicate to ensure that everyone is on the same page with regard to context and vision and motivated to work toward the same shared goals. Here is a list of aspects about which you must regularly communicate with your team:

Long-term goals
: Clearly communicate the overarching vision and mission of the team or project. This helps team members understand the broader purpose behind their work.

Focus areas
: Discuss upcoming priorities and focus areas. This helps the team prepare for challenges and align their efforts accordingly.

Context for tasks
: Relate individual tasks to the larger goals of the team. Providing context helps team members see why and how their contributions matter.

Milestones and achievements
: Regularly update the team on progress, milestones achieved, and notable successes. Celebrating achievements fosters a positive and motivated team culture.

Challenges and roadblocks
: Openly address challenges and roadblocks. This transparency fosters a culture of problem-solving and collaboration.

Changes in strategy or direction
: If there are shifts in strategy or changes in project direction, ensure that the team is promptly informed. Clear communication prevents confusion and aligns everyone with the new direction.

Opportunities for learning and development
: Communicate available learning resources, training, and development opportunities. Encouraging continuous learning contributes to individual and team growth.

Organizational updates
: If there are updates or changes at the organizational level that may impact the team, communicate these changes transparently.

Regular and transparent communication of these facets of a project fosters a cohesive and informed team. It builds trust, ensures everyone is on the same page, and empowers team members to contribute effectively to the team's success.

Structuring for innovation

Inspiration can strike anyone, anywhere. A creative solution or innovative idea can come from anyone on your team, from junior developers to yourself. But, for it to reach fruition, you have to ensure that every idea receives due consideration promptly. For this, you must optimize the team structure, processes, and systems for innovation. This involves the following:

Flatten unnecessary hierarchy
: Create an environment where ideas can flow freely by minimizing unnecessary hierarchical barriers. Encourage team members to share their thoughts and perspectives without fear of retribution.

Informed decisions
: Apple has an organizational structure that centers on functional expertise (*https://oreil.ly/bgqFb*). One of their fundamental beliefs is that "those with the most expertise and experience in a domain should have decision rights for that domain." Putting it in a team perspective, if you want someone to vet a fresh idea, make sure it is someone who has sufficient background in the problem area to make an informed decision.

Emphasize speed
: Innovation often requires agility and rapid adaptation. Reduce bureaucratic processes and empower team members to make decisions quickly, allowing for faster iteration and experimentation.

Adopt minimal viable processes
: Establish minimal viable processes that provide enough structure to guide innovation without stifling creativity. Avoid overplanning and embrace flexibility to adapt to new challenges and opportunities.

Create innovation time
: Set aside dedicated time for team members to work on innovative projects or explore new ideas. This time can be separate from regular project work, allowing for focused creativity.

Facilitate ideation sessions
: Organize brainstorming sessions or workshops where team members can freely share and explore new ideas. This collaborative approach fosters a culture of continuous innovation.

By implementing these strategies, team leaders can create an environment that is conducive to innovation, where creativity is nurtured and team members are empowered to contribute to meaningful advancements within the organization.

Psychological safety

As discussed in Chapter 4, team psychological safety is a critical factor for effective team performance. It is defined as a shared belief that the team is safe for interpersonal risk-taking. As a team leader, you must nurture psychological safety so that team members feel safe speaking up, taking risks, and admitting mistakes.

Creating psychological safety in a tech team setting involves proactive measures by the leader. Great leaders allow unconventional ideas and celebrate failures as learning opportunities. This environment fosters creative thinking without fear of embarrassment. For example, in team meetings, when a member raises a concern about the project, you should openly acknowledge the concern, thank the member for bringing it up, and then facilitate a discussion around it. Ask open-ended questions in such situations to encourage other team members to weigh in, showing that all opinions are valued and important for the team's success.

You can foster psychological safety by encouraging open communication, approaching conflict as a collaborator, and replacing blame with curiosity. This ensures that team members feel comfortable and confident sharing their ideas instead of stifling them. To understand the importance of psychological safety in team settings, consider how a specific scenario plays out.

For example, Ana, a team leader, understands that psychological safety is paramount for effectiveness. During a recent retrospective meeting, Pablo, a normally reserved dev, raised a concern about technical debt accumulation in their codebase.

Ana first publicly thanks Pablo for raising this important issue and acknowledges it as a valid concern impacting velocity and reliability if left unaddressed. She then opens up the discussion by asking nonjudgmental, open-ended questions about the areas of debt and how they affect day-to-day work.

Other engineers start highlighting pain points, realizing it is safe to voice opinions without backlash. Ana listens intently, paraphrasing concerns and recording them on the board. She wraps up by synthesizing the perspectives into potential approaches, while assigning Pablo to lead a task force to produce debt reduction recommendations. In her next team meeting, Ana reviews Pablo's proposals, congratulating him on spearheading this initiative.

Through ongoing reinforcement of positive behaviors, Ana shapes a culture where people feel psychologically safe to participate openly. Had Ana snubbed or ignored Pablo's concerns, she would have missed out on many different perspectives that were shared by the other engineers. Instead, we learn that by proactively thanking members for raising issues, facilitating constructive discussion without judgment, and visibly empowering them to drive improvements, Ana paved the way for an environment where problems can be raised early, without worry of embarrassment or retribution.

Leading diverse teams

I have discussed the relevance of diversity to effective engineering teams in Chapter 1. Diversity is not only a buzzword; it's a necessity for innovation, growth, and long-term success. Diverse teams bring together a wider range of perspectives, experiences, and approaches, leading to more creative solutions, enhanced problem-solving, and a deeper understanding of the diverse customer base that tech companies serve. However, leading diverse teams effectively requires a conscious effort to address unconscious biases, implement inclusive hiring practices, and cultivate an environment that values and celebrates differences. A few helpful strategies for this include the following:

Address unconscious bias

Unconscious bias, the subtle and often unintentional prejudices that influence our thoughts and actions, can hinder the success of diverse teams. To battle this:

- Conduct regular unconscious-bias training for team members to help them recognize and mitigate their own biases.
- Encourage open discussions about unconscious bias and create a safe space for team members to share their experiences.

Promote diverse hiring practices

Diversity starts with the hiring process. To hire a diverse team:

- Expand the reach of recruitment efforts by targeting a wider range of sources, including historically Black colleges and universities, Hispanic-serving institutions, and organizations that support underrepresented groups in tech.
- Partner with organizations that advocate for diversity in tech, such as Women Who Code and Code2040, to gain access to a wider pool of diverse candidates.

Cultivate an inclusive work culture

Fostering an inclusive work environment is crucial for retaining and empowering diverse talent. This means creating a culture where everyone feels valued, respected, and heard:

- Provide opportunities for mentorship and career development for all team members, regardless of their background or experience level.
- Celebrate the unique contributions and perspectives of all team members, recognizing their individual strengths and accomplishments.

As you implement the aforementioned practices, regularly assess the team's progress in terms of diversity, inclusion, and belonging. Seek feedback from team members to identify areas for improvement and make necessary adjustments to policies and practices.

By taking these steps, you can create a more inclusive and equitable workplace where diverse teams can thrive, leading to greater innovation, collaboration, and success for the entire organization.

Identifying potential and developing capability

Effective leadership is not merely about directing tasks and achieving goals; it is about nurturing the potential within each team member and bringing out the best in them. Great leaders are adept at identifying the hidden strengths and motivations of their team members. This enables the leaders to customize goals and responsibilities in a way that aligns with their team members' unique talents and aspirations. A few key points to note when trying to identify and develop talents are as follows:

- Recognize that *talent development is not a one-size-fits-all approach*. Rather, it is a personalized journey that requires a deep understanding of each individual's strengths, weaknesses, and motivations. This journey may involve mentorship, sponsorship, and coaching as fundamental tools to draw out latent potential, enable growth, and transform individuals into high-performing contributors. It can also involve customizing training programs and stretching skills beyond comfort zones.

- Understand that *there is value in small gains*. Incremental progress is the foundation upon which larger achievements are built. Celebrate even the slightest improvements, providing constructive feedback and tailored challenges that encourage individuals to push their boundaries and reach their full potential.
- The art of empowering talent is *not about creating clones of the ideal employee*; it is about cultivating a diverse and vibrant collective of individuals who bring their unique perspectives and strengths to the table.

Feedback is another key component of individual development, providing valuable insights into strengths, areas for improvement, and opportunities for growth. Within software development teams, effective feedback practices foster a culture of continuous learning and improvement, enabling teams to adapt to evolving project needs and achieve their goals.

A structured approach to feedback in software development teams encompasses regular one-on-one meetings, peer reviews, and retrospective meetings. One-on-one meetings provide a dedicated space for individuals to discuss their progress, receive personalized feedback from their managers, and set goals for their continued development. Peer reviews, where team members review each other's work, promote collaborative learning and encourage the exchange of knowledge and expertise. Retrospective meetings, conducted at the end of sprints, offer a collective opportunity for teams to reflect on their performance, identify areas for improvement, and adapt their strategies for future sprints.

This structured approach to feedback ensures that individuals and teams receive regular, constructive feedback that is tailored to their specific needs and the context of the project. By practicing the aforementioned approach, you can cultivate a workforce that is not only capable of achieving extraordinary results but also deeply invested in its own growth and development.

To provide effective feedback, consider the following tips:

Be specific and actionable
: Provide concrete examples and actionable suggestions for improvement. Avoid generalities and focus on behaviors or skills that can be developed.

Balance positive and constructive feedback
: Recognize strengths and achievements while also addressing areas for growth. Use a supportive and encouraging tone to maintain motivation and engagement.

Tailor feedback to the individual
: Consider the unique needs, goals, and learning styles of each team member. Adapt your feedback approach accordingly to ensure maximum impact and receptivity.

Follow up and support
: Regularly check in with team members to discuss progress, provide additional guidance, and celebrate successes. Offer resources and support to help them implement the feedback and continue their development.

Balancing technical expertise with leadership skills

Balancing technical expertise with leadership skills is a crucial challenge for many leaders and requires deliberate effort. While deep technical knowledge is essential for understanding complex problems and making informed decisions, effective leadership requires a broader skill set that encompasses communication, collaboration, and strategic thinking. To excel as a tech leader, it's essential to continuously develop both your technical expertise and your leadership skills. Consider the following strategies:

- Develop technical expertise:
  - Schedule regular periods for updating technical skills, such as coding or learning new technologies.
  - Attend conferences, workshops, and online courses to expand your knowledge base and maintain your technical proficiency.
  - Set aside dedicated time for hands-on technical work or participating in technical workshops.
  - Stay informed about industry trends, emerging technologies, and best practices in software development.
- Enhance leadership skills:
  - Engage in leadership development programs, workshops, or courses to enhance your leadership capabilities.
  - Seek mentorship from experienced leaders within or outside your organization to gain insights and guidance.
  - Regularly reflect on your leadership practices and seek feedback from your team and peers to identify areas for improvement.

- Focus on developing your strategic thinking, decision making, and problem-solving abilities to guide your team toward achieving organizational goals.
- Cultivate strong communication, collaboration, and interpersonal skills to effectively lead and influence others.
- Read books, articles, and case studies on leadership to expand your knowledge and gain new perspectives.

By consistently investing in both technical expertise and leadership skills, you can become a well-rounded and effective tech leader. Remember that skill development is an ongoing process, and there will always be room for growth and improvement. Embrace opportunities to learn, seek out challenges that push you out of your comfort zone, and continuously refine your approach to leadership in the ever-evolving tech industry.

MASTERING THE ATTITUDE

Leading effectively is about walking the talk and requires a lot of self-awareness, adaptability, and a commitment to personal growth. It is about cultivating a mindset that inspires, motivates, and empowers others to achieve their full potential. You must embody the values you want your team to embrace and demonstrate your commitment to the success of the project so that they can follow suit. When leaders lead with conviction, they inspire others to follow and contribute to the collective goals. The key components for achieving this mastery are discussed in the following sections.

Trust and autonomy

In Chapter 1, I discussed how autonomy drives motivation in the case of software engineers. To grant autonomy, you must learn to trust your team members to do the right thing. Exceptional leaders trust their teams, recognizing the intrinsic motivation for quality work. Engineers are not going to self-sabotage their work. Their desire is always to deliver high-quality output, even if their methods for getting there may sometimes be wrong—that is why you need guardrails.

Guardrails, in this context, are the guidelines, processes, and boundaries that you establish to ensure that your team members work effectively and efficiently toward the desired outcomes. These guardrails help prevent potential issues, maintain alignment with project goals, and promote a healthy and productive work environment. Examples of guardrails include the following:

- Clear communication channels and protocols
- Well-defined roles and responsibilities
- Regular check-ins and progress reviews
- Established coding standards and best practices
- Documented decision-making processes

Granting autonomy doesn't mean a lack of oversight; it's about creating an environment of trust and empowerment. Set clear expectations, provide guidance, and establish guardrails to ensure that autonomy is exercised responsibly. This approach not only enhances individual and team morale but also contributes to a culture of innovation and ownership. A few practical ways in which you can grant autonomy are as follows:

Ownership
: Allow teams to take ownership of specific features in a project. Provide them with the autonomy to make decisions regarding the technical approach, timelines, and execution strategies. This fosters a sense of responsibility and creativity.

Flexible work structures
: Grant autonomy in work schedules and arrangements. Recognize that individuals may have different productivity patterns and preferences. Allowing flexible work hours or remote work options empowers team members to manage their time effectively.

Decision making
: Entrust team members with decision-making authority within their areas of expertise. This could include architectural decisions, technology choices, or process improvements. By involving team members in critical decisions, leaders show confidence in their skills and judgment.

Modeling behaviors

Modeling behaviors is a crucial aspect of effective leadership. It involves leading by example and embodying the desired mindsets and values that you want to see reflected in your team members. By consistently demonstrating these behaviors in your day-to-day actions, you signal your priorities and build credibility as a leader. Consistency in your actions reinforces the values and mindsets that you aim to instill within the team. When team members observe their leaders consistently behaving in ways that align with stated values, it builds trust and encourages the team to emulate those behaviors. Here are a few examples of how you can embody specific values or behaviors through your actions:

Demonstrating a growth mindset
: If the organization values a growth mindset and continuous learning, you can embody this by actively seeking opportunities for personal and professional development. This could involve attending training sessions, sharing insights from relevant books or articles, and expressing a willingness to learn from both successes and failures.

Demonstrating inclusiveness
: If fostering a collaborative and inclusive culture is a priority, you must actively participate in team activities and engage openly with everyone on the team. This could happen in both formal and informal settings like team meetings, brainstorming sessions, or a team lunch.

Demonstrating integrity
: If you wish to model integrity, you must demonstrate honesty, trustworthiness, and ethical decision making. Act with transparency, uphold commitments, and take responsibility for actions. Ensure that your behavior aligns with your words—do not make false promises or engage in misleading communication.

A few other behaviors you could model are punctuality, respect, accountability, excellence, positivity, and a commitment to customers.

Making decisions with conviction

Leaders must make difficult choices with conviction and avoid the planning overkill antipattern that I discussed in Chapter 5. Conviction in decision making is essential for effective leadership because it inspires confidence and drives action. When you make decisions with conviction, you project a sense of certainty and resolve that empowers your teams to follow you. This conviction is particularly crucial in uncertain or challenging times, when you need to provide a clear direction and be prepared to own the outcomes as learning opportunities.

A common decision that a team leader must often make is about shipping a product. It is important to know when to stop adding features and ship a product. You must be able to make this decision with conviction, knowing that your team has already done its bit and now it's time to let the users experience it. Let your team members and stakeholders understand that there will be feedback, but that feedback is essential to make the product better. Let your team members get a short respite so that they can be prepared to support the product once it is released.

There will also be times when you may have to change the project plan because of unforeseen challenges. Once you have decided to change the project plan, be prepared to communicate these changes to the team and stakeholders and deal with some resistance.

Data-driven leadership

One of the best ways to avoid making decisions based on random opinions, cognitive bias, or group thinking is by building a culture where everyone on the team uses facts and data to base their decisions. *Data-driven leadership* consists of leading your team based on real-time and accurate data and analytics. In companies with strong data cultures, important decisions are informed by data and analytics and executives act on analytically derived insights rather than intuition or experience.

Software engineering team leaders and managers can leverage data-driven leadership to make better decisions about project planning, resource allocation, risk management, and team development. Data-driven leaders can effectively track project progress, identify potential roadblocks, and make timely course corrections to ensure projects stay on track and meet their objectives. They can also identify areas for improvement in individual and team performance, providing targeted feedback and support to enhance skill development and collaboration.

When we talk about being data driven, metrics and KPIs are often discussed. These are quantifiable measures used to evaluate the success and progress of a team or project. Some common examples in software development include the following:

Velocity
: The amount of work completed by the team in a given sprint or iteration.

Cycle time
: This is the time taken from the initiation of a work item to its completion.

Defect density
: This is the number of defects found per unit of code or per release.

Code coverage
: This is the percentage of code that is executed during automated testing.

Customer satisfaction
: This is measured through surveys or feedback, indicating the level of satisfaction with the delivered product or service.

By establishing clear metrics and KPIs, you can track your team's performance, identify areas for improvement, and make data-driven decisions to optimize your development process.

Key areas where software engineering leaders can employ data-driven practices include the following:

Establishing clear metrics and KPIs
: Define key metrics and KPIs that align with the team's goals and objectives. Track these metrics regularly to assess progress and identify areas for improvement.

Fostering data-driven culture
: Encourage a culture of data-driven decision making by involving team members in data analysis and interpretation.

Communicating data effectively
: Regular reviews of these metrics with the team ensure transparency and collective understanding of project progress and challenges.

Let's look at how data-driven leadership can not only inform but transform teams, with Sandhya's story. An engineering director at Dash Systems, Sandhya has embraced data-driven decision making for continuous improvement. Her team builds Internet of Things (IoT) infrastructure for smart cities. For every new project, Sandhya starts by working with the product leader to define key results aligned to their reliability and scalability goals. This includes performance metrics like system uptime, peak data throughput, and recovery time from outages.

She reviews these indicators during monthly business reviews with stakeholders. Sandhya has also added a dashboard displaying real-time metrics visible to her core team. Trends inform staffing needs—like accelerating hiring after massive data growth. In retrospectives, Sandhya reviews metrics together with the team, celebrating wins and discussing process enhancements.

When velocity dropped an alarming 20% one quarter due to technical debt, Sandhya used accumulated data to convince leaders that a dedicated debt reduction sprint would pay long-term dividends. Hard metrics backed her case.

Within a challenging industry, Sandhya's data-focused leadership has helped maximize her team's impact. The team's metrics-driven approach to resourcing, roadmapping, and process improvements boosts both productivity outcomes and morale.

Adapting to change

In the business of software, like in everything else, change is inevitable, and perhaps it is more likely. There can be many triggers to change—rapid technological advancements, evolving customer and business needs, and unexpected challenges or disruptions can all bring change. As a leader, you must be prepared to adapt to unpredictability, be ready to seek accurate information, and respond quickly with decisive action. For this, it's important to understand the four key principles, also known as the 4 A's of adaptive leadership. The 4 A's, as shown in Figure 7-4, are a framework for understanding and developing the skills necessary to lead effectively in complex and uncertain environments.

Figure 7-4. The 4 A's of adaptive leadership

Anticipation

This is the ability to anticipate future needs, trends, and options. Adaptive leaders are constantly scanning the environment for signals of change and are able to identify potential opportunities and threats before they arise.

Articulation

This is the ability to articulate these needs to build collective understanding and support for action. Adaptive leaders are able to convincingly communicate changes to direction and vision as circumstances change.

Adaptation

This is clearly the core principle and refers to the ability to adapt to changing circumstances. Adaptive leaders are able to adjust their strategies and plans as needed, and they are able to make decisions in the face of uncertainty.

Accountability

This is the ability to include maximum transparency in decision-making processes while taking ownership and responsibility for actions, decisions, and outcomes.

To better understand this framework, let's consider the case of Rajeev, a manager at a software engineering organization. Rajeev was aware that his client, a multinational bank, had been seeking risk-management software that could support its operations in all countries. He recognized that a product his team had already been supporting could be scaled for this purpose. Anticipating a request for proposal, he was prepared to pitch it to the stakeholders when the time came. The proposal was accepted, but the development of an initial MVP was to be fast-paced. Rajeev understood the need for swift action.

The next step was to clearly articulate the vision and strategy to both the existing team and senior management at his organization, securing their agreement. New engineers had to be hired and trained to ensure they could meet the delivery timeline. Rajeev had to foster adaptation within the team, enabling some members to focus on developing the new MVP while others continued to support the existing product while ensuring knowledge transfer between the two teams. This required careful planning and organization to ensure everyone was aligned with the changes and understood their responsibilities. Rajeev effectively demonstrated the 4 A's of adaptive leadership throughout the project, and the team successfully delivered the MVP without major issues within a three-month period.

The 4 A's of adaptive leadership are not mutually exclusive but rather interrelated, each contributing to effective leadership in complex environments. As you develop your skills in each of these areas, you become better equipped to lead your team through change.

Evolving effectiveness into efficiency

In Chapter 2, we talked about becoming effectively efficient. Mastering effective leadership is where your team becomes efficient as a result of sound leadership practices. You can look at this from different angles:

Team
: As a team's capabilities mature, team members become more proficient in their work, and the overall efficiency of the team improves. This can be a result of experience, skill development, and the successful execution of projects.

Processes
: As you identify areas where processes can be streamlined, tools can be optimized, and workflows can be made more efficient, you enhance productivity and output without sacrificing quality. Through failures, you learn to strike a balance between overoptimizing or overstreamlining processes while promoting adaptability and flexibility.

Strategy
: As a result of strategic efforts to enhance effectiveness, resources are utilized more effectively, and more work can be accomplished in less time. You learn to understand the importance of maintaining a balance between operational efficiency and the capacity for creative problem-solving. This ensures that the team remains agile and adaptable in the face of new challenges.

Effective efficiency is a nuanced approach to leadership where the optimization of processes is carefully managed. As capabilities mature, there is an opportunity to enhance efficiency, but you must do so with a mindful eye on maintaining a culture of innovation and adaptability within the team.

In mastering these leadership practices, engineering leaders build elite teams that consistently deliver results and propel the organization forward. Leadership is a continuous learning journey, and great leaders perpetually strive for improvement.

Conclusion

In this concluding chapter, you explored the idea of scaling effectiveness as a leader. You saw how leadership and management are distinct yet complementary roles. Effective leaders inspire and motivate teams to embrace change and achieve organizational goals, while effective managers focus on planning, organizing, and controlling resources to ensure day-to-day operations run smoothly. Leadership and management can be combined by strategic managers who cultivate a visionary approach while also ensuring operational efficiency.

To effectively lead a team, individuals must possess a combination of essential qualities such as technical expertise, communication skills, and agility. Technical expertise enables them to guide their team in solving technical problems effectively. Communication skills are essential for conveying vision, strategies, ideas, and feedback. Agility is essential to adapt to changes.

Leaders should also be able to model personality traits such as self-motivation, drive, integrity, fairness, and humility. Self-motivation fuels the drive to achieve goals, while drive maintains focus on the end goal. Integrity sets the foundation for trust and respect, while fairness ensures equal treatment for all team members. Humility allows leaders to recognize their limitations, learn from mistakes, and foster an approachable environment. By cultivating these qualities, leaders can inspire and motivate their teams to achieve remarkable results.

Effective leaders usually adopt a leadership style and adapt it to their needs. They have a vision and a strategic plan to achieve that vision. They embody the behaviors and qualities that inspire and motivate others and lead with conviction. Effectiveness is a mindset, and the only way you can expand team effectiveness is by embodying effectiveness through your strategy, actions, and attitude.

Index

F

G

M

N

O

P

Q

R

S

T

U

V

W

About the Author

Addy Osmani is a senior staff engineering manager working on Google Chrome. He leads teams focused on making the web fast. For the past 25 years, he has been leading teams in a variety of roles, from an individual contributor mentoring others, through to tech lead, to tech lead manager at varying levels. Passionate about growing the next generation of leaders, Addy has been capturing his notes on what has kept him effective over the years and is sharing them for the first time in this new publication.

Colophon

The cover illustration is original art created by Susan Thompson. The cover fonts are Gilroy Semibold and Guardian Sans. The text fonts are Minion Pro and Scala Pro; the heading and sidebar font is Benton Sans.

Printed in the USA
CPSIA information can be obtained
at www.ICGtesting.com
JSHW012132160724
66518JS00013B/287

9 781098 148249